TEAMWORK IN MULTI-AGENT SYSTEMS

Wiley Series in Agent Technology

Series Editor: Michael Wooldridge, *University of Liverpool, UK*

The 'Wiley Series in Agent Technology' is a series of comprehensive practical guides and cutting-edge research titles on new developments in agent technologies. The series focuses on all aspects of developing agent-based applications, drawing from the Internet, telecommunications, and Artificial Intelligence communities with a strong applications/technologies focus.

The books will provide timely, accurate and reliable information about the state of the art to researchers and developers in the Telecommunications and Computing sectors.

Titles in the series:

Padgham/Winikoff: Developing Intelligent Agent Systems 0-470-86120-7 (June 2004)

Bellifemine/Caire/Greenwood: Developing Multi-Agent Systems with JADE 0-470-05747-5 (February 2007)

Bordini/Hübner/Wooldrige: Programming Multi-Agent Systems in AgentSpeak using Jason 0-470-02900-5 (October 2007)

Nishida: Conversational Informatics: An Engineering Approach 0-470-02699-5 (November 2007)

Jokinen: Constructive Dialogue Modelling: Speech Interaction and Rational Agents 0-470-06026-3 (April 2009)

Castelfranchi/Falcone: Trust Theory: A Socio-Cognitive and Computational Model 0-470-02875-0 (April 2010)

Dunin–Kęplicz/Verbrugge: Teamwork in Multi-Agent Systems: A Formal Approach 0-470-69988-4 (June 2010)

TEAMWORK IN MULTI-AGENT SYSTEMS

A Formal Approach

Barbara Dunin-Kęplicz
Warsaw University and Polish Academy of Sciences
Poland

Rineke Verbrugge
University of Groningen
The Netherlands

⊛WILEY

A John Wiley and Sons, Ltd., Publication

Library of Congress Cataloging-in-Publication Data

Dunin-Keplicz, Barbara.
   Teamwork in multi-agent systems : a formal approach / Barbara Dunin-Keplicz, Rineke Verbrugge.
     p. cm.
   Includes bibliographical references and index.
   ISBN 978-0-470-69988-1 (cloth : alk. paper) 1. Intelligent agents (Computer software) 2. Formal methods (Computer science) 3. Artificial intelligence. I. Verbrugge, Rineke. II. Title.
   QA76.76.I58D98 2010
   006.3 – dc22

                                                2010006086

A catalogue record for this book is available from the British Library.

ISBN 978-0-470-69988-1 (H/B)

Typeset in 10/12 Times by Laserwords Private Limited, Chennai, India.
Printed and Bound in Singapore by Markono Print Media Pte Ltd

To Maksymilian
To Nicole

Contents

About the Authors

Barbara Dunin-Kęplicz

Barbara Dunin-Kęplicz is a Professor of computer science at the Institute of Informatics of Warsaw University and at the Institute of Computer Science of the Polish Academy of Sciences. She obtained her Ph.D. in 1990 on computational linguistics from the Jagiellonian University, and in 2004 she was awarded her habilitation on formal methods in multi-agent systems from the Polish Academy of Sciences.

She is a recognized expert in multi-agent systems. She was one of the pioneers of modeling BDI systems, recently introducing approximate reasoning to the agent-based approach.

Rineke Verbrugge

Rineke Verbrugge is a Professor of logic and cognition at the Institute of Artificial Intelligence of the University of Groningen. She obtained her Ph.D. in 1993 on the logical foundations of arithmetic from the University of Amsterdam, but shortly thereafter moved to the research area of multi-agent systems.

She is a recognized expert in multi-agent systems and one of the leading bridge builders between logic and cognitive science.

Foreword

The ability to cooperate with others is one of the defining characteristics of our species, although of course humans are by no means the only species capable of teamwork. Social insects, such as ants and termites, are perhaps the best-known teamworkers in the animal kingdom and there are many other examples. However, where the human race differs from all other known species is in their ability to apply their teamwork skills to a variety of different domains and to explicitly communicate and reason about teamwork. Human society only exists by virtue of our ability to work together in dynamic and flexible ways. Plus of course, human society exists and functions despite the fact that we all have our own goals, our own beliefs and our own abilities, and in complete contrast to social insects, we are free agents, given fundamental and important control over how we choose to live our lives.

This book investigates teamwork from the point of view of logic. The aim is to develop a formal logical theory that gives an insight into the processes underpinning collaborative effort. The approach is distinguished from related work in for example game theory by the fact that the focus is on the mental states of cooperation participants: their beliefs, desires, and intentions. To be able to express the theory in such terms requires in itself new logical languages, for characterizing the mental state of participants engaged in teamwork. As well as developing the basic model of teamwork, this book explores many surrounding issues, such as the essential link between cooperative action and dialogue.

Michael Wooldridge
University of Liverpool, UK

Preface

The journey of a thousand miles
starts from beneath your feet.

Tao Te Ching (Lao-Tzu, Verse 64)

Teamwork Counts from Two

Barbara and Rineke met at the Vrije Universiteit Amsterdam in the Winter of 1995. The cooperation started blooming as the spring started, mostly during long lasting research sessions in Amsterdam's famous café "De Jaren". Soon Rineke moved to Groningen. Then, on her autumn visits, Barbara survived two floods in Groningen, while Rineke was freezing on her winter trips to Warsaw. Over these years ("de jaren" ...) they started to dream not only about some detachment from their everyday university environment, but especially about a more human-friendly climate when working together. In 2001 Barbara recalled that a place of their dreams exists in reality! Certosa di Pontignano, a meeting place of scholars, situated in the old Carthusian monastery near Siena, Italy, hosted them out of the courtesy of Cristiano Castelfranchi.

Indeed, everything helped them there. A typical Tuscan landscape, commonly considered by visitors as a paradise, the simple, ancient but lively architecture, the amazing beauty of nature, and not to forget: people! Andrea Machetti, Marzia Mazzeschi and their colleagues turned their working visits into fruitful and wonderful experiences. As Barbara and Rineke see it now, the book wouldn't have become real, if Pontignano hadn't been there for them. If one could thank this wonderful place, then they would.

Teamwork Rules

What is contemporary computer science about? Distributed, interactive, autonomous systems are surely in the mainstream, and so are planning and reasoning. These tasks are complex by their very nature, so it is not surprising that in multi-agent environments their complexity tends to explode. Moreover, communication patterns appear to be complex as well. That is where logical modeling is of great help. In this book logic helps us to build minimal, but still workable formal models of teamwork in multi-agent systems. It also lends support when trying to clarify the nature of the phenomena involved, based on the principles of teamwork and other forms of working together, as discovered in the social sciences, management science and psychology. The resulting model TEAMLOG is designed to be lively: to grow or to shrink, but especially to adjust to circumstances when needed. In this logical context, the book is not intended to guide the reader through all possible teamwork-related subjects and the vast multi-disciplinary literature on the subject. It rather presents our personal view on the merits and pitfalls of teamwork in multi-agent settings.

As prerequisites, this book assumes some initial literacy in computer science that students would gain in the first years of a computer science, cognitive science or artificial intelligence curriculum. An introductory course on propositional logic suffices to get a sense of most of the formulas. Some knowledge of modal logic would be helpful to understand the more technical parts, but this is not essential for following the main conceptual line.

As computational agents are the main citizens of this book, we usually refer to a single agent by way of 'it'. If in some example it is clear, on the other hand, that a human agent is meant, we use the conventional reference 'he/she'.

Teamwork Support Matters

First of all, we are grateful to our colleagues who joined our team in cooperative research, leading to articles which later influenced some parts of this book. In particular, we would like to thank Frank Dignum for inspiring collaboration on dialogue – we remember in particular a scientifically fruitful family skiing-and-science trip to Zawoja, Poland. We would also like to thank Alina Strachocka, whose Master's research project under Barbara's wings extended our view on dialogues during collaborative planning. Michał Ślizak, one of Barbara's Ph.D. students, wrote a paper with us on an environmental disaster case study. Finally, Marcin Dziubiński's Ph.D. research under Barbara's supervision led to a number of papers on complexity of teamwork logics.

Discussions with colleagues have found various ways to influence our work. Sometimes a clever member of the audience would point out a counter-example to an early version of our theory. Other times, our interlocutors inspired us with their ideas about dialogue or teamwork. In particular, we would like to thank Alexandru Baltag, Cristiano Castelfranchi, Keith Clark, Rosaria Conte, Frank Dignum, Marcin Dziubiński, Rino Falcone, Wiebe van der Hoek, Erik Krabbe, Theo Kuipers, Emiliano Lorini, Mike Luck, and Andrzej Szałas. Still, there have been many others, unnamed here, to whom we are also indebted.

We gratefully received specially designed illustrations of possible worlds models, team structures and the overarching architecture behind TEAMLOG from Kim Does, Harmen Wassenaar, Alina Strachocka and Andrzej Szałas. In addition, Kim, Michał and Alina also offered a great support by bringing numerous technical tasks to a successful end.

A number of colleagues have generously read and commented various portions of this book. First and foremost, we are very grateful to Andrzej Szałas, who read and suggested improvements on every single chapter! We thank Alina Strachocka, Marcin Dziubiński, Elske van der Vaart, Michał Ślizak and Liliana Pechal for their useful comments on parts of the book. Our students in Groningen and Warsaw, on whom we tried out material in our courses on multi-agent systems, also provided us with inspiring feedback. We would like to thank all of them for their useful suggestions. Any remaining errors are, of course, our own responsibility. Special mention among the students is deserved for

Filip Grządkowski, Michał Modzelewski, and Joanna Zych who inspired some examples of organizational structures in Chapter 4. Violeta Koseska deserves the credit for urging us to write a book together.

From September 2006 through January 2007, Barbara and Rineke worked as Fellows at the Netherlands Institute of Advanced Studies in the Humanities and Social Sciences (NIAS) in Wassenaar. This joint book on teamwork was to be one of the −many!− deliverables of the theme group on *Games, Action and Social Software*, but as is often the case with such projects, the real work of writing and rewriting takes flight afterwards. We would like to thank group co-leader Jan van Eijck for his support. Furthermore, we are grateful to the NIAS staff, in particular to NIAS rector Wim Blockmans and to NIAS head of research planning and support Jos Hooghuis, for their open-mindedness in welcoming our rather unusual project team at NIAS, and for making us feel genuinely at home.

We also highly appreciate the work of our editors at Wiley, Birgit Gruber and Sarah Tilley, for supporting us in the writing process. During the final production process, the book became a real geographically distributed team effort at Wiley, and we would like to thank Anna Smart, Alistair Smith, Shruti Duarah, Jasmine Chang, and David Ando for their contributions.

A number of grants have helped us to work on this book. Both of us would like to acknowledge a NIAS Fellowship. In addition, Barbara would like to acknowlegde the support of the Polish KBN grant 7 T11C 006 20, the Polish MNiSW grant N N206 399334, and the EC grant ALFEBIITE++ (A Logical Framework for Ethical Behaviour between Infohabitants in the Information Trading Economy of the Information Ecosystem, IST-1999-10029). Moreover, Rineke would like to acknowledge the Netherlands Organisation for Scientific Research for three grants, namely NWO ASI 051-04-120 (Cognition Programme Advanced Studies Grant), NWO 400-05-710 (Replacement Grant), and NWO 016-094-603 (Vici Grant).

Finally, we would like to express our immense gratitude to our partners for their steadfast support. Also, we thank them for bearing large part of the sacrifice that goes with such a huge project as writing a book, including having to do without us for long stretches of time.

Barbara Dunin-Kęplicz
Warsaw
keplicz@mimuw.edu.pl

Rineke Verbrugge
Groningen
rineke@ai.rug.nl

1

Teamwork in Multi-Agent Environments

The Master doesn't talk, he acts.
When his work is done,
the people say, 'Amazing:
we did it, all by ourselves!'

Tao Te Ching (Lao-Tzu, Verse 17)

1.1 Autonomous Agents

What is an autonomous agent? Many different definitions have been making the rounds, and the understanding of agency has changed over the years. Finally, the following definition from Jennings *et al.* (1998) has become commonly accepted:

> An agent is a computer system, *situated* in some environment, that is capable of *flexible autonomous* action in order to meet its design objectives.

The environment in which agents operate and interact is usually dynamic and unpredictable.

Multi-agent systems (MASs) are computational systems in which a collection of loosely-coupled autonomous agents interact in order to solve a given problem. As this problem is usually beyond the agents' individual capabilities, agents exploit their ability to *communicate*, *cooperate*, *coordinate* and *negotiate* with one another. Apparently, these complex social interactions depend on the circumstances and may vary from altruistic cooperation through to open conflict. Therefore, in multi-agent systems one of the central issues is the study of how groups work, and how the technology enhancing complex interactions can be implemented. A paradigmatic example of joint activity is *teamwork*, in which a group of autonomous agents choose to work together, both in advancement of their own individual goals as well as for the good of the system as a whole. In the first phase of designing multi-agent systems in the 1980s and 1990s, the emphasis was put on

Teamwork in Multi-Agent Systems: A Formal Approach   Barbara Dunin-Kęplicz and Rineke Verbrugge
© 2010 John Wiley & Sons, Ltd

cooperating teams of software agents. Nowadays there is a growing need for teams consisting of computational agents working hand in hand with humans in *multi-agent environments*. Rescue teams are a good example of combined teams consisting of robots, software agents and people (Sycara and Lewis, 2004).

1.2   Multi-Agent Environments as a Pinnacle of Interdisciplinarity

Variety is the core of multi-agent systems. This simple statement expresses the many dimensions immanent in agency. Apparently, the driving force underlying multi-agent systems is to relax the constraints of the previous generation of complex (distributed) intelligent systems in the field of knowledge-based engineering, which started from expert systems, through various types of knowledge-based systems, up to blackboard systems (Engelmore and Morgan, 1988; Gonzalez and Dankel, 1993; Stefik, 1995). Flexibility is essential for ensuring goal-directed behavior in a dynamic and unpredictable environment. Complex and adaptive patterns of interaction in multi-agent systems, together with agents' autonomy and the social structure of cooperative groups, determine the novelty and strength of the agent-based approach.

Variety is the core of multi-agent systems also because of important links with other disciplines, as witnessed by the following quote from Luck *et al.* (2003):

> A number of areas of philosophy have been influential in agent theory and design. The philosophy of beliefs and intentions, for example, led directly to the BDI model of rational agency, used to represent the internal states of an autonomous agent. Speech act theory, a branch of the philosophy of language, has been used to give semantics to the agent communication language of FIPA. Similarly, argumentation theory – the philosophy of argument and debate, which dates from the work of Aristotle – is now being used by the designers of agent interaction protocols for the design of richer languages, able to support argument and non-deductive reasoning. Issues of trust and obligations in multiagent systems have drawn on philosophical theories of delegation and norms.
>
> Social sciences: Although perhaps less developed than for economics, various links between agent technologies and the social sciences have emerged. Because multiagent systems are comprised of interacting, autonomous entities, issues of organisational design and political theory become important in their design and evaluation. Because prediction of other agents' actions may be important to an agent, sociological and legal theories of norms and group behavior are relevant, along with psychological theories of trust and persuasion. Moreover for agents acting on behalf of others (whether human or not), preference elicitation is an important issue, and so there are emerging links with marketing theory where this subject has been studied for several decades.

1.3   Why Teams of Agents?

Why cooperation?

> Cooperation matters. Many everyday tasks cannot be done at all by a single agent, and many others are done more effectively by multiple agents. Moving a very heavy object is an example of the first sort, and moving a very long (but not heavy) object can be of the second (Grant *et al.*, 2005a).

Teams of agents are defined as follows (Gilbert, 2005):

> The term 'team' tends to evoke, for me, the idea of a social group dedicated to the pursuit
> of a particular, persisting goal: the sports team to winning, perhaps with some proviso as
> to how this comes about, the terrorist cell to carrying out terrorist acts, the workgroup to
> achieving a particular target.

Teamwork may be organized in many different ways. Bratman characterizes shared cooperative activity by the criteria of mutual responsiveness, commitment to joint activity, commitment to mutual support and formation of subplans that mesh with one another (Bratman, 1992). Along with his characteristics, the following essential aspects underlie our approach to teamwork:

- working together to achieve a common goal;
- constantly monitoring the progress of the team effort as a whole;
- helping one another when needed;
- coordinating individual actions so that they do not interfere with one another;
- communicating (partial) successes and failures if necessary for the team to succeed;
- no competition among team members with respect to achieving the common goal.

Teamwork is a highly complex matter, that can be characterized along different lines. One distinction is that teamwork can be primarily defined:

1. In terms of achieving a certain outcome, where the *roles* of agents are of prime importance.
2. In terms of the motivations of agents, where agents' *commitments* are first-class citizens.

In this book, the second point of view is taken.

1.4   The Many Flavors of Cooperation

It is useful to ask initially: what makes teamwork tick? A fair part of this book will be devoted to answering this question.

Coordinated group activity can be investigated from many different perspectives:

- the software engineering perspective (El Fallah-Seghrouchni, 1997; Jennings and Wooldridge, 2000);
- the mathematical perspective (Procaccia and Rosenschein, 2006; Shehory, 2004; Shehory and Kraus, 1998);
- the information theory perspective (Harbers *et al.*, 2008; Sierra and Debenham, 2007);
- the social psychology perspective (Castelfranchi, 1995, 2002; Castelfranchi and Falcone, 1998; Sichman and Conte, 2002);
- the strictly logical perspective (Ågotnes *et al.*, 2008; Goranko and Jamroga, 2004);
- in the context of electronic institutions (Arcos *et al.*, 2005; Dignum, 2006).

We take the *practical reasoning* perspective.

1.5   Agents with Beliefs, Goals and Intentions

Some multi-agent systems are intentional systems implementing *practical reasoning* – the everyday process of deciding, step by step, which action to perform next (Anscombe, 1957; Velleman, 2000). The intentional model of agency originates from Michael Bratman's theory of human rational choice and action (Bratman, 1987). He posits a complex interplay of informational and motivational aspects, constituting together a belief-desire-intention (BDI) model of rational agency.

Intuitively, an agent's *beliefs* correspond to information it has about the environment, including other agents. An agent's *desires* represent states of affairs (options) that it would choose. We usually use the term *goal* for this concept, but for historical reasons we use the abbreviation BDI. In human practical reasoning, *intentions* are first class citizens, as they are not reducible to beliefs and desires (Bratman, 1987). They form a rather special consistent subset of an agent's goals, that it chooses to focus on for the time being. In this way they create a screen of admissibility for the agent's further, possibly long-term, decision process called *deliberation*.

During deliberation, agents decide what state of affairs they want to achieve, based on the interaction of their beliefs, goals and intentions. The next substantial part of practical reasoning is means-ends analysis (or planning), an investigation of actions or complex plans that may best realize agents' intentions. This phase culminates in the construction of the agent's *commitment*, leading directly to action.

In this book, we view software agents from the *intentional stance* introduced by Dennett (1987) as the third level of abstraction (the first two being the physical stance and the design stance, respectively). This means that agents' behavior is explained and predicted by means of mental states such as beliefs, desires, goals, intentions and commitments. The intentional stance, although possibly less accurate in its predictions than the two more concrete stances, allows us to look closer on essential aspects of multi-agent systems. According to Dennett, it does not necessarily presuppose that the agents actually have explicit representations of mental states. In contrast, taking the computer science perspective, we will make agents' mental state representations explicit in our logical framework.

1.6   From Individuals to Groups

A logical model of an agent as an *individual, autonomous* entity has been successfully created, starting from the early 1990s (Cohen and Levesque, 1990; Rao and Georgeff, 1991; Wooldridge, 2000). These systems have been proved to be successful in real-life situations, such as Rao and Georgeff's system OASIS for air traffic control and Jennings and Bussmann's contribution to making Daimler–Chrysler production lines more efficient (Jennings and Bussmann, 2003; Rao and Georgeff, 1995a).

More recently the question how to organize agents' cooperation to allow them to achieve their common goal while striving to preserve their individual autonomy, has been extensively debated. Bacharach notes the following about individual motivations in a team setting (Gold, 2005):

> First, there are questions about motivations. Even if the very concept of a team involves
> a common goal, in real teams individual members often have private interests as well.
> Some individuals may be better motivated than others to 'play for the team' rather than for

themselves. So questions arise for members about whether other members can be trusted to try to do what is best for the team. Here team theory meets *trust* theory, and the currently hot topic of when and why it is rational to trust. Organizational psychology studies how motivations in teams are determined in part by aspects of personality, such as leadership qualities, and by phenomena belonging to the affective dimension, such as mood and 'emotional contagion'.

The intentional stance towards agents has been best reflected in the BDI model of agency. However, even though the BDI model naturally comprises agents' individual beliefs, goals and intentions, these do not suffice for teamwork. When a team is supposed to work together in a planned and coherent way, it needs to present a collective attitude over and above individual ones. Without this, sensible cooperation is impossible, as agents are not properly motivated and organized to act together as a team. Therefore, the existence of collective (or joint) motivational attitudes is a necessary condition for a loosely coupled group of agents to become a strictly cooperative team. As in this book, we focus on cooperation within strictly cooperative teams, cases of competition are explicitly excluded. Strangely enough, many attempts to define coordinated team action and associated group attitudes have neglected the aspect of ruling out competition.

1.7 Group Attitudes

The formalization of informational attitudes derives from a long tradition in philosophy and theoretical computer science. As a result of inspiring discussions in philosophical logic, different axiom systems were introduced to express various properties of the notions of knowledge and belief. The corresponding semantics naturally reflected these properties (Fagin *et al.*, 1995; Hintikka, 1962; Lenzen, 1978). Informational attitudes of groups have been formalized in terms of epistemic logic (Fagin *et al.*, 1995; Meyer and van der Hoek, 1995; Parikh, 2002). Along this line such advanced concepts as general, common and distributed knowledge and belief were thoroughly discussed and precisely defined in terms of agents' individual knowledge or, respectively, belief.

The situation is much more complex in case of motivational attitudes. Creating a conceptually coherent theory is challenging, since bilateral and collective notions cannot be viewed as a straightforward extension or a sort of sum total of individual ones. In order to characterize their collective flavor, additional subtle and diverse aspects of teamwork need to be isolated and then appropriately defined. While this process is far from being trivial, the research presented in this book brings new results in this respect. The complex interplay between environmental and social aspects resulting from the increasing complexity of multi-agent systems significantly contributes to this material. For example, in an attempt to answer what it means for a group of agents to be *collectively committed* to do something, both the circumstances in which the group is acting and properties of the organization it is part of, have to be taken into account. This implies the importance of differentiating the scope and strength of team-related notions. The resulting characteristics may differ significantly, and even become logically incomparable.

1.8 A Logical View on Teamwork: TEAMLOG

Research on a methodology of teamwork for BDI systems led us first to a static, descriptive theory of collective motivational attitudes, called TEAMLOG. It builds on individual goals, beliefs and intentions of cooperating agents, addressing the question what it means for a group of agents to have a *collective intention*, and then a *collective commitment* to achieve a common goal.

While investigating this issue we realized the fundamental role of collective intention in consolidating a group to a strictly cooperating team. In fact, a team is glued together by collective intention, and exists as long as this attitude holds, after which the team may disintegrate. Plan-based collective commitment leads to team action. This plan can be constructed from first principles, or, on the other extreme of a spectrum of possibilities, it may be chosen from a depository of pre-constructed plans. Both notions of collective intentions and collective commitments allow us to express the potential of strictly cooperative teams.

When building a logical model of teamwork, agents' *awareness* about the situation is essential. This notion is understood here as the state of an agent's beliefs about itself, about other agents and about the environment. When constructing collective concepts, we would like to take into account all the circumstances the agents are involved in. Various versions of group notions, based on different levels of awareness, fit different situations, depending on organizational structure, communicative and observational abilities, and so on.

Various epistemic logics and various notions of group information (from distributed belief to common knowledge) are adequate to formalize agents' awareness (Dunin-Kęplicz and Verbrugge, 2004, 2006; Fagin *et al.*, 1995; Parikh, 2002). The (rather strong) notion of *common belief* reflects ideal circumstances, where the communication media operate without failure and delay. Often, though, the environment is less than ideal, allowing only the establishment of weaker notions of group information.

1.9 Teamwork in Times of Change

Multi-agent environments by their very nature are constantly changing:

> As the computing landscape moves from a focus on the individual standalone computer system to a situation in which the real power of computers is realised through distributed, open and dynamic systems, we are faced with new technological challenges and new opportunities. The characteristics of dynamic and open environments in which, for example, heterogeneous systems must interact, span organisational boundaries, and operate effectively within rapidly changing circumstances and with dramatically increasing quantities of available information, suggest that improvements on the traditional computing models and paradigms are required. In particular, the need for some degree of autonomy, to enable components to respond dynamically to changing circumstances while trying to achieve over-arching objectives, is seen by many as fundamental (Luck *et al.*, 2003).

Regardless of the complexity of teamwork, its ultimate goal is always *team action*. Team attitudes underpin this activity, as without them proper cooperation and coordination wouldn't be possible. In TEAMLOG, intentions are viewed as an inspiration for goal-directed activity, reflected in the strongest motivational attitudes, that is in social (or bilateral) and collective commitments. While social commitments are related to individual actions, collective commitments pertain to plan-based team actions.

Basically, team action is nothing more than a coordinated execution of actions from the social plan by agents that have socially committed to do them. The kind of actions is not prescribed: they may vary from basic individual actions like picking up a violin, to more compound ones like carrying a piano, requiring strict coordination of the agents performing them together. In order to start team action, the underlying collective commitment should first be properly constructed in the course of teamwork. Indeed, different individual, social and collective attitudes that constitute the essential components of collective commitment have to be built carefully in a proper sequence. Our approach is based on the four-stage model of Wooldridge and Jennings (1999).

First, during *potential recognition*, an initiator recognizes potential teams that could actually realize the main goal. Then, the proper group is to be selected by him/her and constituted by establishing a collective intention between team members. This takes place during *team formation*. Finally, in the course of *plan formation*, a social plan realizing the goal is devised or chosen, and all agents agree to their shares in it, leading ultimately to collective commitment. At this point the group is ready to start *team action*. When defining these stages we abstract from particular methods and algorithms meant to realize them. Instead, the resulting team attitudes are given.

The explicit model of teamwork provided by TEAMLOG helps the team to monitor its performance and especially to re-plan based on the present situation. The dynamic and unpredictable environment poses the problem that team members may fail to realize their actions or that new favorable opportunities may appear. This leads to the *reconfiguration problem*: how to re-plan properly and efficiently when the situation changes during plan execution? A generic solution of this problem in BDI systems is provided by us in the *reconfiguration algorithm*, showing the phases of construction, maintenance and realization of collective commitment. In fact, the algorithm, formulated in terms of the four stages of teamwork and their complex interplay, is devised to efficiently handle the necessary re-planning, reflected in an *evolution* of collective commitment. Next to the algorithm, the dynamic logic component of TEAMLOG^{dyn} addresses issues pertaining to adjustments in collective commitment during reconfiguration.

The static definitions from TEAMLOG and dynamic properties given in TEAMLOG^{dyn} express solely vital aspects of teamwork, leaving room for case-specific extensions. Under this restriction both parts can be viewed as a set of *teamwork axioms* within a BDI framework. Thus, TEAMLOG formulates postulates to be fulfilled while designing the system. However, one has to realize that *any* multi-agent system has to be tailored to the application in question.

1.10 Our Agents are Planners

"Variety is the core of multi-agent systems." This saying holds also for agents' planning. In early research on multi-agent systems, successful systems such as DMARS, Touring-Machines, PRS and InteRRaP were based on agents with access to plan depositories, from which they only needed to select a plan fitting the current circumstances (d'Inverno *et al.*, 1998; Ferguson, 1992; Georgeff and Lansky, 1987; Müller, 1997). The idea behind this approach was that all possible situations had to be foreseen, and procedures to tackle each of them had to be prepared in advance. These solutions appear to be quite effective in some practical situations. However, over the last few years the time has become ripe for more refined and more flexible solutions.

Taking reconfiguration seriously, agents should be equipped with planning abilities. Therefore our book focuses on the next generation of software agents, who are capable to plan from first principles. They may use contemporary planning techniques such as continual distributed planning (desJardins *et al.*, 1999; Durfee, 2008). Planning capabilities are vital when dealing with real-life complex situations, such as evacuation after ecological disasters. Usually core procedures are pre-defined to handle many similar situations as a matter of routine. However, the environment may change in unpredictable ways that call for time-critical planning as addition to these pre-defined procedures. In such dynamic circumstances, a serious methodological approach to (re-)planning from first principles is necessary. Even so, ubiquitous access to complex planning techniques is still a 'song of the future'.

In this book, we aim to provide the vital methodological underpinnings for teamwork in dynamic environments.

1.11   Temporal or Dynamic?

TEAMLOG has been built incrementally starting from individual intentions, which we view as primitive notions, through social (bilateral) commitments, leading ultimately to collective motivational attitudes. These notions play a crucial role in practical reasoning. As they are formalized in multi-modal logics, their semantics is clear and well defined; this enables us to express many subtle aspects of teamwork like various interactions between agents and their attitudes. The static theory TEAMLOG has been proved sound and complete with respect to its semantics (see Chapter 3 for the proof).

Multi-agent systems only come into their own when viewed in the context of a dynamic environment. Thus, the static logic TEAMLOG is embedded in a richer context reflecting these dynamics. When formally modeling dynamics in logic, the choice is between dynamic logic and temporal logic. Shortly stated, in dynamic logic *actions* (or programs) are first-class citizens, while in temporal logic the *flow of time* is the basic notion (Barringer *et al.*, 1986; Benthem, 1995; Benthem *et al.*, 2006; Doherty and Kvarnström, 2008; Fischer and Ladner, 1979; Fisher, 1994; Harel *et al.*, 2000; Mirkowska and Salwicki, 1987; Salwicki, 1970; Szałas, 1995). Both approaches have their own advantages and disadvantages, as well as proponents and detractors. Lately, the two approaches are starting to be combined and their interrelations are extensively studied, including translations from dynamic presentations into temporal ones (Benthem and Pacuit, 2006). However, the action-related flavor so typical for dynamic logic is hidden in the complex formulas resulting from the translation. Even though the solution is technically satisfying, for modeling applicable multi-agent systems it is appropriate to choose a more recognizable and explicit representation.

We choose agents, actions and plans as the prime movers of our theory, especially in the context of reconfiguration in a dynamic environment. Dynamic logic is eminently suited to represent agents, actions and plans. Thus, we choose dynamic logic on the grounds of clarity and coherence of presentation. Some aspects, such as an agent's commitment strategies, specifying in which circumstances the agent drops its commitments, can be much more naturally formalized in a temporal framework than in a dynamic one. As commitment strategies have been extensively discussed elsewhere (see, for example Dunin-Kęplicz and Verbrugge (1996); Rao and Georgeff (1991)), we shall only informally discuss them in Chapter 4. In addition, the interested reader will find a temporal framework in which our teamwork theory could be embedded in the appendix.

We are agnostic as to which of the two approaches, dynamic or temporal, is better. As Rao and Georgeff did in their version of BDI logic, one can view the semantics of the whole system as based on discrete temporal trees, branching towards the future, where the step to a consecutive node on a branch corresponds to the (successful or failing) execution of an atomic action (Rao and Georgeff, 1991, 1995b). In this view, the states are worlds at a point on a time-branch within a time-tree, so in particular, accessibility relations for individual beliefs, goals and intentions point from such a state to worlds at a (corresponding) point in time.

1.12 From Real-World Data to Teamwork

Formal approaches to multi-agent systems are concerned with equipping software agents with functionalities for reasoning and acting. The starting point of most of the existing approaches is the layer of beliefs, in the case of BDI systems extended by goals and intentions. These attitudes are usually represented in a symbolic, qualitative way. However, one should view this as an idealization. After all, agent attitudes originate from real-world data, gathered by a variety of sources at the *object level* of the system. Mostly, the data is derived from sensors responsible for perception, but also from hardware, different software platforms and last, but not least, from people observing their environment. The point is that this information is inherently quantitative. Therefore one deals with a meta-level duality: sensors provide quantitative characteristics, while reasoning tasks performed at the *meta-level* require the use of symbolic representations and inference mechanisms.

Research in this book is structured along the lines depicted in Figure 1.1. The object-level information is assumed to be summarized in queries returning Boolean values. In this way we will be able to abstract from a variety of formalisms and techniques applicable in the course of reasoning about real-world data. This abstraction is essential, since the

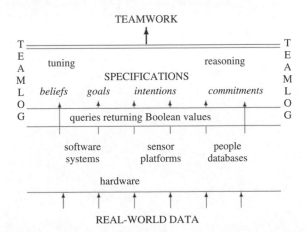

Figure 1.1 The object- and meta-level views on teamwork.

focus of this book is on the *meta-level*, including formal specification and reasoning about teamwork, as exemplified by the static and dynamic parts of TEAMLOG.

1.13   How Complex are Models of Teamwork?

Having a complete static logic TEAMLOG at hand, a natural next step is to investigate the complexity of the satisfiability problem of TEAMLOG, with the focus on individual and collective attitudes up to collective intention. (The addition of collective commitment does not add to the complexity of the satisfiability problem.) Our logics for teamwork are squarely multi-modal, in the sense that different operators are combined and may interfere. One might expect that such a combination is much more complex than the basic multi-agent logic with one operator, but in fact we show that this is not the case. The individual part of TEAMLOG is PSPACE-complete, just like the single modality case. The full system, modeling a subtle interplay between individual and group attitudes, turns out to be EXPTIME-complete, and remains so even when propositional dynamic logic is added to it.

Additionally we make a first step towards restricting the language of TEAMLOG in order to reduce its computational complexity. We study formulas with bounded modal depth and show that in case of the individual part of our logics, we obtain a reduction of the complexity to NPTIME-completeness. We also show that for group attitudes in TEAMLOG the satisfiability problem remains EXPTIME-hard, even when modal depth is bounded by 2. We also study the combination of reducing modal depth and the number of propositional atoms. We show that in both cases this allows for checking the satisfiability of the formulas in linear time.

2

Beliefs in Groups

Not-knowing is true knowledge.
Presuming to know is a disease.
First realize that you are sick;
then you can move toward health.

Tao Te Ching (Lao-Tzu, Verse 71)

2.1 Awareness is a Vital Ingredient of Teamwork

For teamwork to succeed, its participants need to establish a common view on the environment. This can be built by *observation* of both the environment and other agents operating in it, by *communication*, and by *reasoning*. These three important processes pertain to agents' *awareness*. Awareness is understood here as a limited form of consciousness. In the minimal form, it refers to an agent's beliefs about *itself*, about *others* and about *the environment*, corresponding to the *informational stance*. Together they constitute three levels of agents' awareness: *intra-personal* (about the agent itself), *inter-personal* (about other agents as individuals) and *group awareness*.[1]

The research presented in this chapter is meant to contribute to the discussion on formal specifications of agents' awareness in modeling teamwork. Indeed, two issues will be addressed. Firstly, we will argue that agents' awareness becomes a first-class citizen in contemporary multi-agent applications. Secondly, we will point out awareness-related problems. In the subsequent Chapters 3 and 4, we suggest some solutions, implemented in TEAMLOG. The formalization of agents' mental attitudes presented there, constituting a part of a high-level logical specification, are particularly interesting for system

---

[1] This notion of awareness is different than the one used by among others Ågotnes and Alechina (2007b) and Fagin and Halpern (1988). Whereas our notion of awareness refers to an agent's specific informational stance towards a proposition (such as belief or knowledge), their concept of agents *becoming aware* of a proposition denotes that this proposition becomes noticed as relevant by an agent, whether or not it has any belief about its truth value. Fagin *et al.* (1995) give as a possible informal meaning of their awareness formula $A_i\varphi$: 'i is familiar with all propositions mentioned in φ', or alternatively 'i is able to figure out the truth of φ (for example within a given time limit)'. Syntactic approaches to this type of 'relevance awareness' are often used in approaches for modeling agents' awareness by concepts such as *explicit knowledge* (Ågotnes and Alechina, 2007a).

Teamwork in Multi-Agent Systems: A Formal Approach   Barbara Dunin-Kęplicz and Rineke Verbrugge
© 2010 John Wiley & Sons, Ltd

developers when tailoring a multi-agent system for a specific application, especially when both software agents and humans operate in a multi-agent environment. Characterizing the concept of awareness, we aim to forge synergy between the cognitive science and multi-agent systems perspectives and results. Importantly, cognitive science analyzes and explains problems of human awareness, which can be successfully and precisely translated into the BDI framework. Then, resulting solutions may be easily compared and formally verified. In this way, the two fields mutually benefit from each other's point of view.

This chapter is structured as follows. In Section 2.2, we shortly describe the different ways in which agents' awareness in dynamic environments can be created, including their possible pitfalls. Section 2.3 gives the logical background about the modal logics used in this chapter, including the language and possible worlds semantics. Then, in Sections 2.4 and 2.5, we choose well-known axiom systems for beliefs and knowledge, respectively, treating the properties of individual and group notions of awareness. Section 2.6 describes some difficulties when combining knowledge and belief in a single logical system. Section 2.7 forms the heart of this chapter, delineating problems concerning agents' awareness in multi-agent environments. Subsection 2.7.1 focuses on agents' awareness about their own mental states and the effects of their bounded rationality. In Subsection 2.7.2, attention is given to problematic aspects of agents' models of other individuals' mental states. These strands come together in Subsection 2.7.3 where we show that awareness in groups is of vital importance. We discuss some pitfalls in achieving it and point to the next chapters presenting some possibilities for system developers to flexibly adapt the type of group awareness in a multi-agent system to the environment and the envisioned kind of organization.

2.2 Perception and Beliefs

Agents' awareness builds on various forms of observation, communication and reasoning. In multi-agent systems awareness is typically expressed in terms of *beliefs*. One may ask: why belief and not knowledge?

The concept of *knowledge* usually covers more than a true belief (Artemov, 2008; Lenzen, 1978). In fact, an agent should be able to justify its knowledge, for example by a proof. Unfortunately, in the majority of multi-agent system applications, such justification cannot be guaranteed. The reasons for this, are complex. It is perception that provides the main background for agents' informational stance. However, the natural features of perception do not lead to optimism:

- limited accuracy of sensors and other devices;
- time restrictions on completing measurements;
- unfortunate combinations and unpredictability of environmental conditions;
- noise, limited reliability and failure of physical devices.

In real systems, this imprecise, incomplete and noisy information of a quantitative nature resulting from perception should be filtered and intelligently transformed into a qualitative presentation. This rather difficult step is the subject of ongoing research on approximate multi-agent environments (Doherty *et al.*, 2003, 2007; Dunin-Kęplicz and Szałas, 2007, 2010). An interesting problem in this research is fusion of approximate information from heterogeneous agents with different abilities of perception (Dunin-Kęplicz *et al.*, 2009a). Apparently, the information resulting from this process cannot be proved to be *always true*, something that is taken for granted in the case of knowledge. Finally, computational limits of perception may give rise to false beliefs or to beliefs that, while true, still cannot be justified by the agent. Therefore, a standard solution accepted in agency is to express the results of an agent's perception in terms of beliefs.

Furthermore, agents' beliefs are naturally *communicated* to others by means of dialogues or communication protocols. However, communication channels may be of uncertain quality, so even if a trustworthy sender knows a certain fact, the receiver may only believe it. Finally, agents' *reasoning*, under natural computational limits, sometimes may lead to false conclusions. Despite these pessimistic characteristics of beliefs reflecting a pragmatic view on agency, in the sequel, we will keep the idealistic assumption that an agent's beliefs are at least consistent.

2.3 Language and Models for Beliefs

As mentioned before, we propose the use of modal logics to formalize agents' informational attitudes. Table 2.1 below gives the formulas appearing in this chapter, together with their intended meanings. The symbol φ denotes a proposition.

2.3.1 The Logical Language for Beliefs

Formulas are defined with respect to a fixed finite set of agents. The basis of the inductive definition is given in the following definition.

Definition 2.1 (Language) The language is based on the following two sets:

- a denumerable (finite or infinite) set $\mathcal{P}$ of *propositional symbols*;
- a finite set $\mathcal{A}$ of *agents*, denoted by numerals $1, 2, \ldots, n$.

$\mathcal{P}$ and $\mathcal{A}$ are disjoint.

Definition 2.2 (Formulas) We inductively define a set $\mathcal{L}$ of formulas as follows.

F1 each atomic proposition $p \in \mathcal{P}$ is a formula;
F2 if φ and ψ are formulas, then so are $\neg\varphi$ and $\varphi \wedge \psi$;
F3 if φ is a formula, $i \in \mathcal{A}$, and $G \subseteq \mathcal{A}$, then the following **epistemic modalities** are formulas: $\text{BEL}(i, \varphi)$; $\text{E-BEL}_G(\varphi)$; $\text{C-BEL}_G(\varphi)$.

The constructs $\top$, $\bot$, $\vee$, $\rightarrow$ and $\leftrightarrow$ are defined in the usual way, as follows:

- $\top$ abbreviates $\neg(p \wedge \neg p)$ for some atom $p \in \mathcal{P}$;

Table 2.1 Formulas and their intended meanings.

$\text{BEL}(i, \varphi)$	Agent i believes that φ
$\text{E-BEL}_G(\varphi)$	Group G has the general belief that φ
$\text{C-BEL}_G(\varphi)$	Group G has the common belief that φ

- $\bot$ abbreviates $p \land \neg p$ for some atom $p \in \mathcal{P}$;
- $\varphi \lor \psi$ abbreviates $\neg(\neg\varphi \land \neg\psi)$;
- $\varphi \to \psi$ abbreviates $\neg(\varphi \land \neg\psi)$;
- $\varphi \leftrightarrow \psi$ abbreviates $\neg(\varphi \land \neg\psi) \land \neg(\psi \land \neg\varphi)$.

2.3.2   Kripke Models for Beliefs

Each Kripke model for the language $\mathcal{L}$ consists of a set of worlds, a set of accessibility relations between worlds and a valuation of the propositional atoms, as follows.

Definition 2.3 (Kripke model)   A Kripke model is a tuple $\mathcal{M} = (W, \{B_i : i \in \mathcal{A}\}, Val)$, such that:

1. W is a non-empty set of possible worlds, or states;
2. For all $i \in \mathcal{A}$, it holds that $B_i \subseteq W \times W$. They stand for the accessibility relations for each agent with respect to *beliefs*. $(s, t) \in B_i$ means that t is an 'epistemic alternative' for agent i in state s;[2] henceforth, we often use the notation $s B_i t$ to abbreviate $(s, t) \in B_i$;
3. $Val : (\mathcal{P} \times W) \to \{0, 1\}$ is the function that assigns truth values to ordered pairs of atomic propositions and states (where 0 stands for false and 1 for true).

In the possible worlds semantics above, the accessibility relations B_i lead from worlds w to 'epistemic alternatives': worlds that are consistent with agent i's beliefs in w. Thus, the meaning of BEL can be defined informally as follows: agent i believes φ (BEL(i, φ)) in world w, if and only if, φ is true in all agent i's epistemic alternatives with respect to w. This is reflected in the formal truth definition 2.4. The definition above places no constraints on the accessibility relations. In Section 2.4, we will show how certain restrictions on the accessibility relations correspond to natural properties of beliefs.

At this stage, it is possible to define the truth conditions pertaining to the language $\mathcal{L}$, as far as the propositional connectives and individual modal operators are concerned. The expression $\mathcal{M}, s \models \varphi$ is read as 'formula φ is satisfied by world s in structure $\mathcal{M}$'.

Definition 2.4 (Truth definition)

- $\mathcal{M}, s \models p \Leftrightarrow Val(p, s) = 1$, where $p \in \mathcal{P}$;
- $\mathcal{M}, s \models \neg\varphi \Leftrightarrow \mathcal{M}, s \not\models \varphi$;
- $\mathcal{M}, s \models \varphi \land \psi \Leftrightarrow \mathcal{M}, s \models \varphi$ and $\mathcal{M}, s \models \psi$;
- $\mathcal{M}, s \models$ BEL(i, φ) iff $\mathcal{M}, t \models \varphi$ for all t such that $s B_i t$.

2.4   Axioms for Beliefs

To represent beliefs, we adopt a standard KD45$_n$-system for n agents as explained in Fagin *et al.* (1995) and Meyer and van der Hoek (1995), where we take BEL(i, φ) to have as an intended meaning 'agent i believes proposition φ'.

---

[2] For beliefs, in the literature often the term 'doxastic' is used instead of 'epistemic'.

2.4.1   Individual Beliefs

KD45$_n$ consists of the following axioms and rules for $i = 1, \ldots, n$:

A1   All instantiations of propositional tautologies

A2$_B$ $\text{BEL}(i, \varphi) \wedge \text{BEL}(i, \varphi \to \psi) \to \text{BEL}(i, \psi)$ (Belief Distribution)

A4$_B$ $\text{BEL}(i, \varphi) \to \text{BEL}(i, \text{BEL}(i, \varphi))$ (Positive Introspection)

A5$_B$ $\neg \text{BEL}(i, \varphi) \to \text{BEL}(i, \neg \text{BEL}(i, \varphi))$ (Negative Introspection)

A6$_B$ $\neg \text{BEL}(i, \perp)$ (Belief Consistency)

R1   From φ and $\varphi \to \psi$ infer ψ (Modus Ponens)

R2$_B$ From φ infer $\text{BEL}(i, \varphi)$ (Belief Generalization)

Note that there is no axiom **A3** here: in analogy to our later axiom system for the logic of knowledge, **A3** would refer to the truth principle $\text{BEL}(i, \varphi) \to \varphi$, which is highly implausible for a logic of belief. In the system KD45$_n$, axiom **A3** is replaced by the weaker Belief Consistency axiom **A6**.

The name KD45$_n$ derives from the history of modal logic. In the classical publication by Lemmon (1977), axiom **A2** has been named **K** and principle **A3** has been named **T**. Already in Lewis and Langford (1959), **A4** was referred to as **4** and **A5** as **5**. Later, axiom **A6** has been named **D**. Thus, different systems are named for different combinations of axioms, followed by a subscript for the number of agents.

Note that one can apply the Generalization rule **R2$_B$** only to formulas that have been proved already, thus to *theorems* of KD45$_n$, and not to formulas that depend on assumptions. After all, $\varphi \to \text{BEL}(i, \varphi)$ is not a valid principle.

As usual in modal logic, it is fruitful to look for correspondences between the axiom system and the semantics. The following relations are well-known (Blackburn *et al.*, 2002; van Benthem, 2005):

Positive Introspection A4$_B$ corresponds to transitivity;

Negative Introspection A5$_B$ corresponds to Euclidicity;

Belief Consistency A6$_B$ corresponds to seriality.

Therefore, on the basis of our choice of axioms, we follow the tradition in epistemic logic by supposing the accessibility relations B_i to be:

transitive: $\forall w_1, w_2, w_3 \in W \; (w_1 B_i w_2 \text{ and } w_2 B_i w_3 \Rightarrow w_1 B_i w_3)$;

Euclidean: $\forall w_1, w_2, w_3 \in W \; (w_1 B_i w_2 \text{ and } w_1 B_i w_3 \Rightarrow w_2 B_i w_3)$;

serial: $\forall w_1 \exists w_2 \; w_1 B_i w_2$.

Note that, in the semantics, the accessibility relations B_i need not be reflexive, corresponding to the fact that an agent's beliefs need not be true (See Figure 2.1 for a typical KD45$_n$ model.).

It has been proved that KD45$_n$ is sound and complete with respect to these semantics (Fagin *et al.*, 1995; Meyer and van der Hoek, 1995).

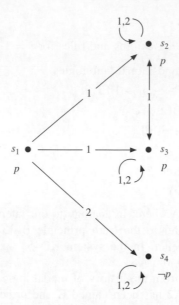

Figure 2.1 Typical KD45$_n$ model with accessibility relations B_i (represented by arrows labeled with the respective agent names). The accessibility relations are transitive, serial and Euclidean, but not reflexive. The following hold: $\mathcal{M}, s_1 \models \text{BEL}(1, p)$ and $\mathcal{M}, s_1 \models \text{BEL}(2, \neg p)$; however, also $\mathcal{M}, s_1 \models \text{BEL}(1, \text{BEL}(2, p))$ and $\mathcal{M}, s_1 \models \text{BEL}(2, \text{BEL}(1, \neg p))$, so the agents are mistaken in their second-order beliefs about p.

2.4.2 From General to Common Belief

When building a logical model of teamwork, there is a strong need for different types of group beliefs, as explained before (see Figure 2.2). Indeed, one can define various modal operators for group beliefs. The formula E-BEL$_G(\varphi)$, called 'general belief in φ', is meant to stand for 'every agent in group G believes φ'. It is defined semantically as:

$$\mathcal{M}, s \models \text{E-BEL}_G(\varphi) \text{ iff for all } i \in G, \ \mathcal{M}, s \models \text{BEL}(i, \varphi)$$

which corresponds to the following axiom:

C1 E-BEL$_G(\varphi) \leftrightarrow \bigwedge_{i \in G} \text{BEL}(i, \varphi)$ (General Belief)

A traditional way of lifting single-agent concepts to multi-agent ones is through the use of *common belief* C-BEL$_G(\varphi)$. This rather strong operator is similar to the more usual one of common knowledge, except that a common belief among a group that φ need not imply that φ is true.

C-BEL$_G(\varphi)$ is meant to be true if everyone in G believes φ, everyone in G believes that everyone in G believes φ, etc. Let E-BEL$_G^1(\varphi)$ be an abbreviation for E-BEL$_G(\varphi)$ and let E-BEL$_G^{k+1}(\varphi)$ for $k \geq 1$ be an abbreviation of E-BEL$_G(\text{E-BEL}_G^k(\varphi))$. Thus we have $\mathcal{M}, s \models \text{C-BEL}_G(\varphi)$ iff $\mathcal{M}, s \models \text{E-BEL}_G^k(\varphi)$ for all $k \geq 1$.

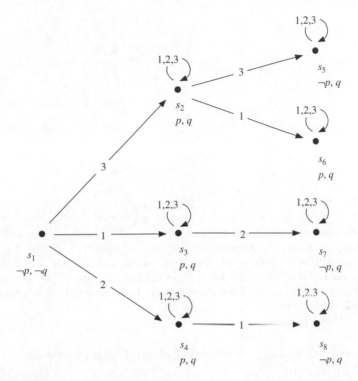

Figure 2.2 For general and common beliefs for group $G = \{1, 2, 3\}$, we have $\mathcal{M}, s_1 \models \text{E-BEL}_G(p)$ but not $\mathcal{M}, s_1 \models \text{E-BEL}^2_G(p)$, for instance because s_5 is accessible from s_1 in two steps by accessibility relations for agents 2 and 3, respectively, and $\mathcal{M}, s_5 \models \neg p$. Therefore, it is not the case that $\mathcal{M}, s_1 \models \text{C-BEL}_G(p)$. On the other hand, q holds in all worlds that are G_B-reachable from s_1, namely $\mathcal{M}, s_i \models q$ for $s_i \in \{s_2, s_3, s_4, s_5, s_6, s_7, s_8\}$. Therefore $\mathcal{M}, s_1 \models \text{C-BEL}_G(q)$. Notice that $\mathcal{M}, s_1 \not\models q$, so the group has a 'common illusion' about q.

Define world t to be G_B-*reachable* from world s iff $(s, t) \in (\bigcup_{i \in G} B_i)^+$, the transitive closure of the union of all individual accessibility relations. Formulated more informally, this means that there is a path of length ≥ 1 in the Kripke model from s to t along accessibility arrows B_i that are associated with members i of G. Then the following property holds (see Fagin *et al.* (1995)):

$$\mathcal{M}, s \models \text{C-BEL}_G(\varphi) \text{ iff } \mathcal{M}, t \models \varphi \text{ for all } t \text{ that are } G_B\text{-reachable from } s$$

Using this property, it can be shown that the following axiom and rule can be soundly added to the union of KD45_n and **C1**:

C2 $\quad \text{C-BEL}_G(\varphi) \leftrightarrow \text{E-BEL}_G(\varphi \wedge \text{C-BEL}_G(\varphi))$ $\qquad\qquad$ (Common Belief)

RC1 $\quad$ From $\varphi \rightarrow \text{E-BEL}_G(\psi \wedge \varphi)$ infer $\varphi \rightarrow \text{C-BEL}_G(\psi)$ $\qquad\qquad$ (Induction Rule)

The resulting system is called KD45^C_n, and is sound and complete with respect to Kripke models where all n accessibility relations are transitive, serial and Euclidean (Fagin *et al.*, 1995). The following useful theorem and rule can easily be derived from KD45^C_n:

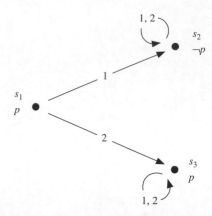

Figure 2.3 A counter-example against collective negative introspection: a KD45$_n$ model with accessibility relations B_i (represented by arrows labeled with the respective agent names). The accessibility relations are transitive, serial and Euclidean, but not reflexive. The following holds for $G = \{1, 2\}$: $\mathcal{M}, s_1 \models \neg\text{C-BEL}_G(p)$ because $\mathcal{M}, s_2 \models \neg p$; however, $\mathcal{M}, s_1 \not\models {}^\sim\text{C-BEL}_G(\neg\text{C-BEL}_G(p))$. Even stronger, there is a false belief about the common belief: $\mathcal{M}, s_1 \models \text{BEL}(2, \text{C-BEL}_G(p))$.

C3 $\text{C-BEL}_G(\varphi) \wedge \text{C-BEL}_G(\varphi \to \psi) \to \text{C-BEL}_G(\psi)$ (Common Belief Distribution)

RC2 From φ infer $\text{C-BEL}_G(\varphi)$ (Common Belief Generalization Rule)

In the sequel, we will also use the following standard properties of C-BEL$_G$ (see, for example, Fagin *et al.* (1995, Exercise 3.11)).

Lemma 2.1 *Let* $G \subseteq \{1, \ldots, n\}$ *be given. Then the following hold for all formulas* φ, ψ:

- $\text{C-BEL}_G(\varphi \wedge \psi) \leftrightarrow \text{C-BEL}_G(\varphi) \wedge \text{C-BEL}_G(\psi)$ (Conjunction Distribution)
- $\text{C-BEL}_G(\varphi) \to \text{C-BEL}_G(\text{C-BEL}_G(\varphi))$ (Collective Positive Introspection)

Remark 2.1 Note that we do not have negative introspection for common beliefs: it may be the case that something is not commonly believed in a group, but the group is not aware of this lack of common belief! See Figure 2.3 for a counter-example.

2.5 Axioms for Knowledge

Knowledge, which always corresponds to the facts and can be justified by a formal proof or less rigorous argumentation, is the strongest individual informational attitude considered in this book.

In order to represent knowledge, we take $\text{KNOW}(i, \varphi)$ to have as intended meaning 'agent i knows proposition φ' (see Table 2.2). Next, we adopt the standard S5$_n$-system for n agents as explained in Fagin *et al.* (1995) and Meyer and van der Hoek (1995), containing the following axioms and rules for $i = 1, \ldots, n$. They are similar to the axioms

Table 2.2   Formulas and their intended meanings.

$KNOW(i, \varphi)$	Agent i knows that φ
$E\text{-}KNOW_G(\varphi)$	Group G has the general knowledge that φ
$C\text{-}KNOW_G(\varphi)$	Group G has the common knowledge that φ

for belief, except that **A6**$_B$ is replaced by the stronger **A3**$_K$. In addition to **A1** and **R1**, here follow the axioms and rule for knowledge:

A2$_K$ $KNOW(i, \varphi) \wedge KNOW(i, \varphi \to \psi) \to KNOW(i, \psi)$ (Knowledge Distribution)

A3$_K$ $KNOW(i, \varphi) \to \varphi$ (Veracity of Knowledge)

A4$_K$ $KNOW(i, \varphi) \to KNOW(i, KNOW(i, \varphi))$ (Positive Introspection)

A5$_K$ $\neg KNOW(i, \varphi) \to KNOW(i, \neg KNOW(i, \varphi))$ (Negative Introspection)

R2$_K$ From φ infer $KNOW(i, \varphi)$ (Knowledge Generalization)

Here, the names of the axioms have historical origins in modal logic, similarly to those for beliefs, thus the axiom system could have been called KT45$_n$ in analogy to KD45$_n$. However, for this most well-known axiom system for epistemic logic we keep to the historical name S5, which was already introduced in Lewis and Langford (1959).

Just as for beliefs, we can introduce operators for group knowledge, starting with *general knowledge* $E\text{-}KNOW_G(\varphi)$, which stands for 'everyone knows φ'.

The strongest notion of knowledge in a group is *common knowledge* $C\text{-}KNOW_G(\varphi)$, which is the basis of all conventions and the preferred basis of coordination (Lewis, 1969).

The operators for group knowledge obey axioms similar to those for general and common belief, with the addition of a truth axiom:

CK1 $E\text{-}KNOW_G(\varphi) \leftrightarrow \bigwedge_{i \in G} KNOW(i, \varphi)$ (General Knowledge)

CK2 $C\text{-}KNOW_G(\varphi) \leftrightarrow E\text{-}KNOW_G(\varphi \wedge C\text{-}KNOW_G(\varphi))$ (Common Knowledge)

CK3 $C\text{-}KNOW_G(\varphi) \to \varphi$ (Truth of Common Knowledge)

RCK1   From $\varphi \to E\text{-}KNOW_G(\psi \wedge \varphi)$ infer   (Induction Rule)
            $\varphi \to C\text{-}KNOW_G(\psi)$

This results in the well-known system S5$_n^C$, which is sound and complete with respect to models in which the accessibility relations for knowledge are equivalence relations (that is reflexive, transitive and symmetric) (Fagin *et al.*, 1995). The following useful theorem and rule can easily be derived from S5$_n^C$:

CK4 $C\text{-}KNOW_G(\varphi) \wedge C\text{-}KNOW_G$ (Common Knowledge Distribution)
            $(\varphi \to \psi) \to C\text{-}KNOW_G(\psi)$

RCK2   From φ infer $C\text{-}KNOW_G(\varphi)$ (Common Knowledge Generalization Rule)

A very positive feature of common knowledge is that if C-KNOW$_G$ holds for ψ, then C-KNOW$_G$ also holds for all logical consequences of ψ. The same is true for common belief. Thus, in the axiom systems for the relevant epistemic languages we have *belief and knowledge distribution rules*. For the proofs, we use the modal axioms and rules mentioned below plus propositional logic.

Lemma 2.2 *The following derivation rules can be proved in the systems which have the relevant operators in their language:*

- $\vdash \psi \rightarrow \chi \Rightarrow \vdash \text{BEL}(i, \psi) \rightarrow \text{BEL}(i, \chi)$ *(by* $\mathbf{R2}_B$ *and* $\mathbf{A2}_B$*)*
- $\vdash \psi \rightarrow \chi \Rightarrow \vdash \text{KNOW}(i, \psi) \rightarrow \text{KNOW}(i, \chi)$ *(by* $\mathbf{R2}_K$ *and* $\mathbf{A2}_K$*)*
- $\vdash \psi \rightarrow \chi \Rightarrow \vdash \text{E-BEL}_G^k(\psi) \rightarrow \text{E-BEL}_G^k(\chi)$ *(by* $\mathbf{R2}_B$, $\mathbf{C1}$, *and* $\mathbf{A2}_B$, *k-fold)*
  for all $k \geq 1$
- $\vdash \psi \rightarrow \chi \Rightarrow \vdash \text{E-KNOW}_G^k(\psi) \rightarrow$ *(by* $\mathbf{R2}_K$, $\mathbf{CK1}$, *and* $\mathbf{A2}_K$, *k-fold)*
  $\text{E-KNOW}_G^k(\chi)$ *for all* $k \geq 1$
- $\vdash \psi \rightarrow \chi \Rightarrow \vdash \text{C-BEL}_G(\psi) \rightarrow \text{C-BEL}_G(\chi)$ *(by* $\mathbf{RC2}$ *and* $\mathbf{C3}$*)*
- $\vdash \psi \rightarrow \chi \Rightarrow \vdash \text{C-KNOW}_G(\psi) \rightarrow \text{C-KNOW}_G(\chi)$ *(by* $\mathbf{RCK2}$ *and* $\mathbf{C4}$*)*

Thus, agents reason in a similar way from ψ and commonly believe in this similar reasoning and the final conclusions. This property is crucial when modeling teamwork, as it ensures that agents build models of others in a coherent way.

2.6   Relations between Knowledge and Belief

As mentioned before, it is not the case that knowledge is the same as true belief, that is, it does *not* hold that:

$$\text{KNOW}(i, \varphi) \leftrightarrow \text{BEL}(i, \varphi) \wedge \varphi$$

There are interesting discussions in the literature about whether knowledge is the same as *justified* true belief or whether it should even be something stronger (Artemov, 2008; Gettier, 1963).

How should one combine knowledge and belief into a single logical system? Kraus and Lehmann (1988) introduced a system now called KL_m, an apparently simple combination of S5_m for the K-operators and KD45_m for the B-operators, with as only additions the following two mixed axioms:

$$\mathbf{KB1} \quad K_i\varphi \rightarrow B_i\varphi$$

$$\mathbf{KB2} \quad B_i\varphi \rightarrow K_i B_i\varphi$$

Problem

It appears that in KL_m, an agent cannot believe to know a false proposition, namely: $\text{KL}_m \vdash B_i K_i\varphi \rightarrow K_i\varphi$, and therefore by $\mathbf{A3}_K$, an undesired consequence results:

$$\text{KL}_m \vdash B_i K_i\varphi \rightarrow \varphi.$$

Proof sketch for $\text{KL}_2 \vdash B_1 K_1 p \rightarrow K_1 p$:

1. $\text{KL}_2 \vdash B_1 K_1 p \rightarrow \neg B_1 \neg K_1 p$
   (by $\mathbf{A6}_B$ in its form $\neg(B_i\varphi \wedge B_i\neg\varphi)$, plus propositional logic).
2. $\text{KL}_2 \vdash \neg B_1 \neg K_1 p \rightarrow \neg K_1 \neg K_1 p$
   (**KB1** and propositional logic: contraposition).

3. $KL_2 \vdash \neg K_1 \neg K_1 p \rightarrow K_1 p$

   ($\mathbf{A5}_K$ and propositional logic: contraposition).
4. $KL_2 \vdash B_1 K_1 p \rightarrow K_1 p$

   (from 1,2,3 by propositional logic: hypothetical syllogism).

This means, combining $KL_2 \vdash B_1 K_1 p \rightarrow K_1 p$ with **KB1** and **KB2** and positive intro-spection, that knowledge reduces to belief: $KL_2 \vdash B_1 p \leftrightarrow K_1 p$.

Some authors think that **KB2** is the culprit. In contrast, Halpern (1996) argues that **KB1** is questionable, and that one should only assume knowledge to imply belief ($K_i \varphi \rightarrow B_i \varphi$) for factual formulas φ, without any modal operators.

General problems when combining two modal systems (such as the 'blow-up' of complexity as shown by Blackburn and Spaan (1993)) are often due to the fact that axioms are schemas, applicable to all formulas in the (combined) language. This also appears to create the problem in the mixed system KL_m. In fact, Halpern's solution circumvents this problem by using a restricted language. In Chapter 9, we will refer to more general results showing that restricting the language of the full logic of teamwork, the multi-modal theory TEAMLOG, may also lead to lower complexity. For more on combining knowledge and belief, see Voorbraak (1991).

2.7 Levels of Agents' Awareness

When creating a framework that could be used in mixed teams composed of software agents, robots and people, we need to take agents' limits into account, including human *bounded rationality*. According to Herbert Simon, who coined the term bounded rationality, 'boundedly rational agents experience limits in formulating and solving complex problems and in processing (receiving, storing, retrieving, transmitting) information' (Williamson, 1981). We agree with Simon that models which present humans as logically omniscient or as perfectly rational in the sense of optimizing their own utility are problematic. We extend this discussion to software agents, which also need to reason under bounded rationality, as they operate under time and other resource constraints. Let us investigate the cognitive limits on the three levels of human and software agents' awareness: *intra-personal* (about the agent itself), *inter-personal* (about other agents as individuals) and *group awareness*.

2.7.1 Intra-Personal Awareness

Intra-personal awareness or consciousness of one's own mental states, also called meta-consciousness, plays an important role in an agent's thinking and reasoning. Such intro-spection has for long been considered as totally unproblematic:

> Consciousness was often viewed as though it was *the* defining feature of human thought. The philosophical traditions that have had the strongest influence on psychology are those of Locke and Descartes, and while these two didn't agree on much, the one proposition they shared was that cognitive states are transparent to introspection (Litman and Reber, 2005).

In the second half of the 20th Century, however, cognitive scientists started to study phenomena like implicit cognition. It appeared that experimental subjects could correctly recognize well-formed strings of abstract languages by learning from examples, without being able to formulate the complex underlying rule (Litman and Reber, 2005). Thus, humans are often not aware of their own knowledge and beliefs.[3] In this section, we will see how the epistemic logics that are usually used in multi-agent systems to model agents' knowledge and belief do not in general form an accurate model of human cognitive abilities.

2.7.1.1 Problems of Logical Omniscience

A first problem for modeling human agents is that, as mentioned above, they often lack positive or negative introspection into their own knowledge and beliefs. As a counter-example to negative introspection, one may be completely unaware of a sentence φ that one doesn't believe, and thus not believe that one does not believe it. A counter-example to positive introspection is formed by the implicit cognition experiments mentioned above. In multi-agent systems, in order for agents to model themselves properly, the developer needs to take care that a modicum of (especially positive) introspection is present.

Another problem of systems based on epistemic logic is that we have the following theorems (similar ones hold for knowledge instead of belief):

$$\models \varphi \Rightarrow \ \models BEL(i, \varphi) \qquad\qquad\qquad \text{(Belief of Valid Formulas)}$$
$$\models \varphi \to \psi \Rightarrow \ \models BEL(i, \varphi) \to BEL(i, \psi) \qquad \text{(Closure under Valid Implication)}$$

These are examples of *logical omniscience*: agents believe all theorems, as well as all logical consequences of their beliefs. Any modal logic with standard Kripke semantics in which belief is formalized as a necessity operator has this property. Logical omniscience definitely does not apply to people, nor to software agents, who have only limited time available. It is unrealistic to assume that they believe every logical theorem, however complicated.

Two belief-related problems of logical omniscience are:

$$BEL(i, \varphi) \to \neg BEL(i, \neg\varphi) \qquad\qquad\qquad \text{(Consistency of Beliefs)}$$
$$BEL(i, (BEL(i, \varphi) \to \varphi)) \qquad\qquad \text{(Belief of Having no False Beliefs)}$$

The first one is problematic because the agent may believe two sentences which are in fact (equivalent to) each other's negation without the agent being aware of it. The second one (which follows from $A6_B$ and $A5_B$ from Subsection 2.4.1) makes an agent too idealistic about its beliefs: it is not aware of its own limitations.

There are several possible solutions to the problems of logical omniscience, involving non-standard semantics or syntactic operators for awareness and explicit belief. Good logical references to the logical omniscience problem and its possible solutions are Meyer and van der Hoek (1995, Chapter 2) and Fagin *et al.* (1995, Chapter 9). Interesting recent views on logical omniscience can be found in Alechina *et al.* (2006), Parikh (2005) and Roy (2006).

---

[3] For interesting recent research on the importance of unconscious deliberation, see Dijksterhuis *et al.* (2006).

2.7.2 Inter-Personal Awareness

Bounded rationality plays a role not only in limiting intra-personal awareness but it also constrains agents' reasoning about other agents' mental states.

Formal models of human reasoning, such as those in epistemic logic and game theory, assume that humans can faultlessly reason about other people's individual knowledge and beliefs, for example in card games such as bridge and happy families (van der Hoek and Verbrugge, 2002). However, recent research in cognitive psychology reveals that adults do not always correctly use their theory of what others know in concrete situations (Flobbe et al., 2008; Hedden and Zhang, 2002; Keysar et al. 2003; Verbrugge and Mol, 2008).

In Keysar's experiments, some adult subjects could not correctly reason in a practical situation about another person's lack of knowledge (first-order theory of mind reasoning of the form 'a does not know p') (Keysar et al., 2003). Hedden and Zhang (2002), when describing their experiments involving a sequence of dyadic games, suggested that players generally began with first-order reasoning. When playing against first-order co-players, some began to use second-order reasoning (for example, of the form 'a does not know that I know that p'), but most of them remained on the first level (Hedden and Zhang, 2002).[4]

In recent experiments by Verbrugge and Mol (2008), it turns out that many humans *can* play a version of symmetric Mastermind involving natural language utterances such as 'some colors are right' reasonably well. The successful experimental subjects develop a winning strategy for the game by using a higher-order theory of mind: 'Which sentences reveal the least information while still being true?', 'What does the opponent think I am trying to make him think?' (Mol et al., 2005). Thus, they apply their awareness of others' mental states about them in a new practical situation.

2.7.2.1 Inter-Personal Awareness in BDI Systems

Using axioms $\mathbf{A3}_K$, $\mathbf{A2}_K$ and rule $\mathbf{R2}_K$, one can derive:

$$\text{KNOW}(i, \text{KNOW}(j, \varphi)) \rightarrow \text{KNOW}(i, \varphi) \qquad \text{(Transparency)}$$

However, this is not realistic for people. A child may know that her father has proved Fermat's last theorem, without knowing the theorem herself (where 'knowing' includes being able to justify it). This so-called transparency problem has been treated by using an alternative semantics of 'local worlds' (Gochet and Gillet, 1991).

Finally, the following theorem follows from $\mathbf{A6}_B$ and $\mathbf{R2}_B$:

$$\text{BEL}(j, \text{BEL}(i, \varphi)) \rightarrow \text{BEL}(j, \neg\text{BEL}(i, \neg\varphi))$$

This is also unrealistic: people sometimes have no idea whether their friends are consistent in their beliefs or not. Summing up, the above and similar theorems about epistemic operators presuppose that agents are constantly aware that other agents follow the epistemic rules, for example by monitoring the consistency of their beliefs. It even presupposes that agents believe that others are in their turn aware of still other agents following the logical rules. These types of 'transparency' of logical omniscience are certainly not plausible for human beings.

---

[4] This may actually be an effect of Hedden and Zhang's training sessions, which seemed to suggest a first-order strategy would always be successful, so that subjects had to 'unlearn' this strategy during the test phase of the experiment (Flobbe et al., 2008).

2.7.3   Group Awareness

If even limited orders of theory of mind, as in inter-personal awareness, present such
difficulties for humans, it seems that creating group awareness is impossible: reasoning
about common belief and common knowledge apparently involves an infinitude of levels.
From the time when these notions were first studied, there has been a puzzle about their
establishment and assessment, the so-called *Mutual Knowledge Paradox*, most poignantly
described in Clark and Marshall (1981). How can it be that to check whether one makes a
felicitous reference when saying 'Have you seen the movie showing at the Roxy tonight?',
one has to check an infinitude of facts about reciprocal knowledge, but people seem to
do this in an instant. Clark's solution for human communication was that such *common
ground* (common knowledge) about a sentence can be created if a number of conditions is
met, namely co-presence, mutual visibility, mutual audibility, co-temporality, simultaneity,
sequentiality, reviewability and revisability. Most of these conditions do not hold in multi-
agent systems, where agents communicate over non-instantaneous and possibly faulty
communication media.

The most pressing problem with common knowledge is that it is hard or impossible
to attain in situations where the communication channel is not commonly known to be
trustworthy. Halpern and Moses (1992) proved that common knowledge of certain facts is
on the one hand necessary for coordination in well-known standard examples, while on the
other hand, it cannot be established by communication if there is any uncertainty about
the communication channel (Fagin *et al.*, 1995). More concretely, in file transmission
protocols at any time only a bounded level of knowledge E-KNOW$_G^{k+1}(\varphi)$ (and belief as
well) about the message is achieved (Halpern and Zuck, 1987; Stulp and Verbrugge, 2002).
Good references to the difficulties concerning the attainment of common knowledge, as
well as to possible solutions, are given is Fagin *et al.* (1995, Chapter 11).

Even though common knowledge cannot in general be established by communication,
we have shown that common belief can, in very restricted circumstances. It is possible to
give a procedure that can, under some very strong assumptions about the communication
channels, trust by group members of the initiator and temporary persistence of some
relevant beliefs, establish common beliefs.

More specifically, suppose an initiator a wants to establish C-BEL$_G(\varphi)$ within a fixed
group $G = \{1, \ldots, n\}$, where $a \in G$. Informally and from a higher-level view, the procedure works by the initiator a sending messages as follows:

1. a sends the message φ to agents $\{1, \ldots, n\}$ in an interleaved fashion, where each
   separate message is sent from a to i using the alternating-bit protocol or TCP;
2. Then in the same way, a sends the message C-BEL$_G(\varphi)$ to agents $\{1, \ldots, n\}$;
3. Recipients send acknowledgements of bits received (as by the alternating-bit protocol
   and TCP) but need not acknowledge the receipt of the full message.

Now suppose that the communication channel is fair and allows only one kind of error
(from deletion, mutation and insertion), and that all agents trust the initiator with respect to
its announcements. Then finally, all agents believe the messages they receive from a, and
a believes them as well. Thus, after all agents have received the messages, we will have:

$$\mathrm{BEL}(i, \varphi \wedge \mathrm{C\text{-}BEL}_G(\varphi))$$

for all $i \leq n$; thus by axiom **C1**, we have:

$$\text{E-BEL}_G(\varphi \wedge \text{C-BEL}_G(\varphi))$$

which by axiom **C2** is equivalent to $\text{C-BEL}_G(\varphi)$, as desired.

Notice that the reason that this procedure can establish common belief, whereas common knowledge can never be established, is exactly that common beliefs need not be true. Thus, initiator a may believe and utter $\text{C-BEL}_G(\varphi)$ even if $\text{C-BEL}_G(\varphi)$ has not yet, in fact, be established. Thus, if φ is in fact true, $\varphi \wedge \text{C-BEL}_G(\varphi)$ is an example of the belief-analogue of a 'successful formula' as defined in dynamic epistemic logic, namely a formula that comes to be commonly believed by being publicly announced (van Ditmarsch and Kooi, 2003).

The problem with this procedure is that it only works under the assumption of a kind of 'blind trust' of the group members in the initiator's message. If the communication medium is noisy and one of the other agents i reasons about this and about the initiator's beliefs about the noisiness, then at the moment just after receiving the message, i may believe that the initiator still believes that i has not yet received his message.[5] This concern can be overcome if one makes a much stronger assumption on the communication medium, namely that there are no real errors and that there is a maximum delay in reception of messages sent, which is commonly believed by all agents (similarly to commonly known delays in Fagin *et al.* (1995)). In most real situations, one would not want to restrict to agents overly trusting in authorities or to very safe communication media.

Alternatively, in a much wider set of circumstances, one can suitably apply communication protocols that establish ever higher approximations of common knowledge or common belief within a group (Brzezinski *et al.*, 2005; Van Baars and Verbrugge, 2007). These protocols are less efficient because the needed number of messages passed back and forth between the initiator and the others growing linearly in the desired level k of group knowledge or belief ($\text{E-KNOW}_G^k(\psi)$ respectively $\text{E-BEL}_G^k(\psi)$), but the assumptions are much easier to guarantee.

We acknowledge that issues about the possibility of establishing common belief and common knowledge are important and should be adequately solved. For the next few chapters, however, we focus on the formalization of collective motivational attitudes needed for teamwork, and for the time being we choose to base it on the relatively simple logic for common belief, characterized by the axiom system KD45_n^C defined above. Thus, even though we view common belief as an idealization, it is still a good *abstraction tool* to study teamwork.

2.7.4 Degrees of Beliefs in a Group

It is well known that for teamwork, as well as coordination, it often does not suffice that a group of agents has a general belief to a certain proposition ($\text{E-BEL}_G(\psi)$), but they should commonly believe it ($\text{C-BEL}_G(\psi)$). For example, in team actions like lifting a heavy object together or *coordinated attack*, the success of each individual agent and their mutual coordination are vital to the overall result:

---

[5] This concern was voiced to us by Emiliano Lorini (personal communication).

Two divisions of an army are camped on two hilltops overlooking a common valley. In the valley awaits the enemy. It is clear that if both divisions attack the enemy simultaneously they will win the battle, whereas if only one division attacks it will be defeated. The divisions do not initially have plans for launching an attack on the enemy, and the commanding general of the first division wishes to coordinate a simultaneous attack (at some time the next day). Neither general will decide to attack unless he is sure that the other will attack with him. The generals can only communicate by means of a messenger. Normally, it takes the messenger one hour to get from one encampment to the other. However, it is possible that he will get lost in the dark or, worse yet, be captured by the enemy. Fortunately, on this particular night, everything goes smoothly. How long will it take them to coordinate an attack? (Halpern and Moses, 1984).

It has been proved that for such an attack to be guaranteed to succeed, the starting time of the attack must be a common belief (even common knowledge) for the generals involved (Fagin *et al.*, 1995). In short, one could say that common knowledge and common belief are hard to achieve, even though they express natural concepts. When investigating the level of group awareness necessary in teamwork, we realized that this issue has to be studied in detail each and every time when tailoring a multi-agent system for a specific application. The system developer should be very careful when it comes to deciding about an effective, but still minimal level of group beliefs. Indeed, in some situations general belief suffices perfectly well, while at some other time it needs to be iterated a couple of times, and in other contexts still, a full-fledged common belief is required.

Clearly there exists a vast spectrum of possibilities, for example between individual knowledge or belief and the collective informational attitudes. For example, establishing general belief places much less constraints on the communication medium and the communication protocol than common belief does. This subject has been investigated by Parikh and Krasucki (1992) and Parikh (2002). In fact, they introduced a hierarchy of levels of knowledge in terms of individual knowledge and common knowledge for different subgroups, and proved a number of interesting mathematical properties.

Their definition of levels is based on the notion of *embeddability* of finite strings of individual knowledge operators into one another: the string *aba* is embeddable in itself, in *aaba* and in *abca* (notation *aba* ≤ *abca*), but not in *aabb*. Parikh and Krasucki (1992) extend this notion to so-called *C-embeddability* for the epistemic language with common knowledge operators for groups. They stipulate that in addition to the normal embeddability conditions, C-KNOW$_G$ ≤ C-KNOW$_{G'}$ if $G \subseteq G'$.

After introducing these definitions of embeddability, they show that the resulting order is a well partial order, which means that it is well-founded and that every set of mutually incomparable elements is finite. Moreover, they prove that for all histories and all strings over the language, if $x \leq y$ and in some world yp is true, then so is xp; Parikh and Krasucki use the notation KNOW$_i \varphi$, which corresponds to KNOW(i, φ) used in this book. For example, if:

$$\mathcal{M}, s \models \text{KNOW}_i \text{C-KNOW}_{\{i,j,k\}} K_j p$$

then also:

$$\mathcal{M}, s \models \text{C-KNOW}_{\{j,k\}} \text{KNOW}_j p$$

because indeed:

$$\text{C-KNOW}_{\{j,k\}}\text{KNOW}_j \leq \text{KNOW}_i\text{C-KNOW}_{\{i,j,k\}}\text{KNOW}_j.$$

Levels of knowledge thus correspond to downwardly closed sets with respect to $\leq$. Parikh and Krasucki (1992) show that there are only countably many levels of knowledge, which are all recognizable by finite automata. It turns out, however, that the similarly defined hierarchy based on individual beliefs and common beliefs for different subsets of agents is structurally different from the knowledge hierarchy, due to the lack of the truth axiom. In particular, they prove that there are uncountably many levels of belief (Parikh, 2002).

Returning to the practice of teamwork, the final decision about the level k of iteration of general belief $(\text{E-BEL}_G^k(\psi))$ in a specific context and related to the application in question, each and every time hinges on determining a good balance between communication and reasoning. This problem will be further investigated in the next chapters.

3

Collective Intentions

A good traveler has no fixed plans
and is not intent upon arriving.

Tao Te Ching (Lao-Tzu, Verse 27)

3.1 Intentions in Practical Reasoning

When investigating collective *motivational* notions, the concept of a group of agents is essential. This book focuses on a specific kind of group, namely a team, defined in Weiss (1999) as follows:

> A *team* is a group in which the agents are restricted to having a common goal of some sort. Typically, team members cooperate and assist each other in achieving their common goal.

In a similar vein, Wooldridge and Jennings (1999) point to the vital role of intentions in teamwork:

> The key mental states that control agent behavior in our model are intentions and joint intentions – the former define local asocial behavior, the latter control social behavior. Intentions are so central because they provide both the stability and predictability that is necessary for social interaction, and the flexibility and reactivity that is necessary to cope with the changing environment.

Practical reasoning is the form of reasoning that is aimed at conduct rather than knowledge (see also Section 1.5). The cycle of this reasoning involves:

- repeatedly updating beliefs about the environment;
- deciding what options are available;
- 'filtering' these options to determine new intentions;
- creating commitments on the basis of intentions;
- performing actions in accordance with commitments.

Teamwork in Multi-Agent Systems: A Formal Approach   Barbara Dunin-Kęplicz and Rineke Verbrugge
© 2010 John Wiley & Sons, Ltd

Practical reasoning involves two important processes: deciding *what* goals need to be achieved and then *how* to achieve them. The former process is known as *deliberation*, the latter as *means-end reasoning*. In the sequel we will discuss them in the context of informational and motivational attitudes of the agents involved in teamwork.

The key concept in the theory of practical reasoning is the one of *intention*, studied in-depth in Bratman (1987). Intentions form a rather special consistent subset of goals, that the agent wants to focus on for the time being. According to Cohen and Levesque (1990), intention consists of choice together with commitment (in a non-technical sense). In our approach these two ingredients are separated: an intention is viewed as a chosen goal, providing inspiration for a more concrete social (pairwise) commitment in the individual case, and a plan-based collective commitment in the group case.

In Jacques Ferber's book on multi-agent systems (Ferber, 1999), intention is character-ized from the psychological viewpoint:

> The concept of intention was (and still is) one of the most controversial in the history of psychology. Certain people – the eliminativists – purely and simply refuse to introduce this concept into their theories, claiming not only that it is useless, but also that it mindlessly con-fuses the issues. Others, in contrast, think of it as one of the essential concepts of psychology and that it should be given a central role, for it constitutes a keystone of the explanation of human behaviour in terms of mental states. Finally, the psycho-analytical school sees it as merely a vague concept which is handy in certain cases, but which should generally be replaced by desire and drives, which alone are capable of taking account of the overall behaviour of the human being in his or her aspirations and suffering.

We do not aim to present a psychologically sound theory of motivations driving human behaviour. Instead we study motivational aspects of *rational* decision making, disregard-ing irrational drives and desires, which make human behavior difficult to interpret and hard to predict. In our analysis we do not consider any specific notion of *rationality*, such as the economic one used in game theory. We solely assume that agents are logical reasoners.

There is a common agreement that intentions play a number of important roles in practical reasoning, such as the following, summarized from the seminal work of Bratman (1987) and Cohen and Levesque (1990):

I1 *Intentions drive means-end-reasoning.*
I2 *Intentions constrain future deliberation.*
I3 *Intentions persist long enough, according to a reconsideration strategy.*
I4 *Intentions influence beliefs upon which future practical reasoning is based.*

Thus intentions create a screen of admissibility for the agent's further, possibly long-term, deliberation. However, from time to time intentions should be reconsidered, due to the dynamics of the situation. For example they are achieved already, they will never be achieved or reasons originally supporting them hold no longer. This requires balancing *pro-active*, (that is goal-directed) and *reactive* (that is event-driven) behavior. Indeed, in this book we carefully maintain this balance on the three levels of teamwork: individual, social and collective.

On the individual and social level the problem of persistence of both intentions and then commitments is first expressed in an agent's *intention* and *commitment strategies*, addressing the question: *when and how can an agent responsibly drop its intention or commitment?* The answer to this question is discussed in Chapter 4, but see also the influential paper on BDI architectures by Rao and Georgeff (1991), as well as Dunin-Kęplicz and Verbrugge (1996, 1999).

3.1.1 Moving Intentions to the Collective Level

The intuition behind group intention is undoubtedly more advanced but also somehow mysterious to interpret. We definitely refrain from investigating the psychological flavor of this notion. Instead we focus on the instrumental aspects that allow a team to cooperate smoothly. While the mechanisms behind group intentionality might stay hidden in the course of psychological analysis, logical modeling requires isolating them and enhancing their transparency.

In our approach, collective intention not only integrates the team, but also helps to monitor teamwork. Essentially, to speak about collective forms of intentions in truly cooperative teams, all group members need to share both a common goal as well as individual intention towards this goal. Clearly this is not sufficient for attitude revision. To adjust to changing circumstances, even if some members drop their individual intentions, the team re-plans, aiming to ultimately realize the collective intention. What exactly is needed over and above team members' individual intentions forms the main question driving the current chapter.

In the sequel, we present our way of understanding collective intentions, together with examples of situations where they apply. In contrast to many other approaches to intentions in a group, we provide a completeness proof for the logic TEAMLOG with respect to the intended semantics. The system is known to be EXPTIME-complete, so in general it is not feasible to give automated proofs of desired properties. At least there is no single algorithm that performs well on all inputs. As with other modal logics, the better option would be to develop a variety of different algorithms and heuristics, each performing well on a limited class of inputs. For example, it is known that restricting the number of propositional atoms to be used or the depth of modal nesting may reduce the complexity (cf. Graedel (1999), Halpern (1995), Hustadt and Schmidt (1997), Vardi (1997) and Chapter 9). Also, when considering specific applications it is possible to reduce some of the infinitary character of common beliefs and collective intentions to more manageable proportions (cf. Fagin *et al.* (1995, Chapter 11) and Chapters 4 and 7). We will extensively discuss complexity issues of teamwork logics in Chapter 9. In this chapter we leave out temporal and dynamic considerations (see Section 1.11 for a discussion of possible choices). The presented definitions of collective intentions in terms of more basic attitudes may be combined with either choice, depending on the application.

The rest of this chapter is structured in the following manner. Section 3.2 gives a short logical background. Section 3.3 gives a description of the logical theory of individual goals and intentions. The role of collective intention in teamwork is informally discussed in Section 3.4. The heart of the chapter is formed by Section 3.5, in which collective intentions are investigated and characterized in a logical framework. Next, Section 3.6 presents several approaches to handling the potentially infinitary flavor of collective intentions. Some alternative definitions of collective intention, based on the notion of degrees of awareness, are presented in Section 3.7. The completeness of the logic of mutual intention and its nearest neighbors is proved in Section 3.8. The chapter rounds off with a discussion of related work.

3.2 Language and Models for Goals and Intentions

Tables 3.1 and 3.2 below give the new formulas appearing in this chapter (in addition to those of Table 2.1), together with their intended meanings. The symbol φ denotes a proposition.

3.2.1 The Logical Language

We extend the logical language $\mathcal{L}$ of Chapter 2 by adding a clause for motivational attitudes to the inductive step of Definition 2.2:

Definition 3.1 (Formulas) We inductively define a set of formulas as in Definition 2.2, with the additional clause:

F4 if φ is a formula, $i \in \mathcal{A}$, and $G \subseteq \mathcal{A}$, then the following are formulas:
  motivational modalities GOAL(i, φ), INT(i, φ), E-INT$_G(\varphi)$
    M-INT$_G(\varphi)$, M-INT$'_G(\varphi)$, C-INT$_G(\varphi)$, C-INT$'_G(\varphi)$

3.2.2 Kripke Models

The Kripke models of Definition 2.3 are extended with accessibility relations for motivational attitudes, as follows:

Definition 3.2 (Kripke model) A Kripke model is a tuple
  $\mathcal{M} = (W, \{B_i : i \in \mathcal{A}\}, \{G_i : i \in \mathcal{A}\}, \{I_i : i \in \mathcal{A}\}, Val)$, such that:

1. W is a set of possible worlds, or states;
2. For all $i \in \mathcal{A}$, it holds that $B_i, G_i, I_i \subseteq W \times W$. They stand for the accessibility relations for each agent with respect to *beliefs*, *goals* and *intentions*, respectively. For example, $(s, t) \in I_i$ means that t is an alternative state consistent with agent i's intentions in state s. Henceforth, similarly as for beliefs, we often use the notation $sG_i t$ to abbreviate $(s, t) \in G_i$, and $sI_i t$ to abbreviate $(s, t) \in I_i$;
3. $Val : \mathcal{P} \times W \to \{0, 1\}$ is the function that assigns the truth values to ordered pairs of atomic propositions and states.

At this stage, it is possible to extend the truth conditions pertaining to the language $\mathcal{L}$ of Definition 2.4 with conditions for individual motivational operators. As a reminder, the expression $\mathcal{M}, s \models \varphi$ is read as 'formula φ is satisfied by world s in structure $\mathcal{M}$'.

Definition 3.3 (Truth definition)

- $\mathcal{M}, s \models$ GOAL(i, φ) iff $\mathcal{M}, t \models \varphi$ for all t such that $sG_i t$.
- $\mathcal{M}, s \models$ INT(i, φ) iff $\mathcal{M}, t \models \varphi$ for all t such that $sI_i t$.

3.3   Goals and Intentions of Individual Agents

TEAMLOG is minimal in the sense of dealing solely with the most substantial aspects of teamwork. Additional elements appearing on the stage in specific cases may be addressed by refining the system and adding new axioms. This subsection focuses on individual goals and intentions, and gives a short overview of our choice of axioms (adapted from Rao and Georgeff (1991)) and the corresponding semantic conditions (see Table 3.1 for the formulas). In this chapter, we leave out considerations of the aspects of time and actions in order to focus on the main problem, the definition of collective intentions in terms of more basic attitudes.

Table 3.1   Individual formulas and their intended meanings.

$GOAL(a, \varphi)$	Agent a has the goal to achieve φ
$INT(a, \varphi)$	Agent a has the intention to achieve φ

In this section we delineate the individual part of TEAMLOG, called TEAMLOG^{ind}. These theories officially carry a subscript n for the number of agents in $\mathcal{A}$, but for a given situation n is fixed so in the sequel the subscript is usually suppressed.

TEAMLOG^{ind} includes the axioms for individual beliefs. For the motivational operators GOAL and INT the axioms include the system K, which we adapt for n agents to K_n. For $i = 1, \ldots, n$ the following axioms and rules are included:

A1   All instantiations of tautologies of the
            propositional calculus

A2$_G$ $GOAL(i, \varphi) \wedge GOAL(i, \varphi \rightarrow \psi) \rightarrow GOAL(i, \psi)$ (Goal Distribution)

A2$_I$ $INT(i, \varphi) \wedge INT(i, \varphi \rightarrow \psi) \rightarrow INT(i, \psi)$ (Intention Distribution)

R1   From φ and $\varphi \rightarrow \psi$ infer ψ (Modus Ponens)

R2$_G$ From φ infer $GOAL(i, \varphi)$ (Goal Generalization)

R2$_I$ From φ infer $INT(i, \varphi)$ (Intention Generalization)

In a BDI system, an agent's activity starts from goals. In general, the agent may have many different objectives which will not all be pursued. As opposed to intentions, goals do not directly lead to actions, so an agent can behave rationally, even though it has different inconsistent goals. Thus, in contrast to Rao and Georgeff (1991), who assumed a goal consistency axiom, we restricted ourselves to the basic system K_n for goals. Then, the agent chooses some goals to become intentions. Without going into details on intention adoption (but see Chapter 8, Dignum and Conte (1997) and Dignum et al. (2001b)), we assume that intentions are chosen in such a way that consistency is preserved. Thus for intentions we assume, as Rao and Georgeff (1991) do, that they should be consistent:

A6$_I$ $\neg INT(i, \bot)$ for $i = 1, \ldots, n$ (Intention Consistency Axiom)

Note that Axiom **A6$_I$** is logically equivalent to $INT(i, \varphi) \rightarrow \neg INT(i, \neg\varphi)$, the scheme **D** known from modal logic (Blackburn et al., 2002).

Intentions do lead to action, but when the intended proposition is already satisfied, this may sometimes be the empty action. For example, by Rule **R$_I$**, all propositional tautologies are intended by every agent.

It is not hard to prove soundness and completeness of the basic axiom systems for goals and intentions with respect to suitable classes of models by a tableau method, and also give decidability results using a small model theorem. See Chapter 9 for the needed proof methods. These are applied to the more complicated combined system that also includes interdependency axioms to relate the informational and motivational attitudes. We turn to this combined system now.

3.3.1   Interdependencies between Attitudes

Mixed axioms of two kinds are added to the basic system. The first kind expresses introspection properties, relating motivational attitudes to beliefs about them, while the second kind relates goals to intentions.

3.3.1.1   Introspection about Goals and Intentions

Interdependencies between belief and individual motivational attitudes are expressed by the following axioms for $i = 1, \ldots, n$:

$\mathbf{A7}_{GB}$	$\mathrm{GOAL}(i, \varphi) \rightarrow \mathrm{BEL}(i, \mathrm{GOAL}(i, \varphi))$	(Positive Introspection for Goals)
$\mathbf{A7}_{IB}$	$\mathrm{INT}(i, \varphi) \rightarrow \mathrm{BEL}(i, \mathrm{INT}(i, \varphi))$	(Positive Introspection for Intentions)
$\mathbf{A8}_{GB}$	$\neg\mathrm{GOAL}(i, \varphi) \rightarrow \mathrm{BEL}(i, \neg\mathrm{GOAL}(i, \varphi))$	(Negative Introspection for Goals)
$\mathbf{A8}_{IB}$	$\neg\mathrm{INT}(i, \varphi) \rightarrow \mathrm{BEL}(i, \neg\mathrm{INT}(i, \varphi))$	(Negative Introspection for Intentions)

These four axioms express that agents are aware of the goals and intentions they have, as well as of the lack of those that they do not have. Notice that we do not add the axioms of *strong realism* that Rao and Georgeff adopt for a specific set of formulas φ, the so-called O-formulas: $\mathrm{GOAL}(i, \varphi) \rightarrow \mathrm{BEL}(i, \varphi)$ and $\mathrm{INT}(i, \varphi) \rightarrow \mathrm{BEL}(i, \varphi)$, corresponding to the fact that an agent believes that it can optionally achieve its goals and intentions by carefully choosing its actions. These axioms correspond to semantic restrictions on the branching time models considered in Rao and Georgeff (1991).[1]

Also, we do not adopt the converse axiom of *realism* advocated by Cohen and Levesque (1990): $\mathrm{BEL}(i, \varphi) \rightarrow \mathrm{GOAL}(i, \varphi)$. In their formalism, where a possible world corresponds to a time line, the realism axiom expresses that agents adopt as goals the inevitable facts about the world.

3.3.1.2   Fact

The semantic property corresponding to $\mathbf{A7}_{IB}$ is:

$$\forall s, t, u((s B_i t \wedge t I_i u) \Rightarrow s I_i u)$$

while analogously, $\mathbf{A7}_{GB}$ corresponds to:

$$\forall s, t, u((s B_i t \wedge t G_i u) \Rightarrow s G_i u).$$

---

[1] Because of the restriction to O-formulas, both these versions of realism are intimately connected to the choice of temporal structure, a question that we leave out of consideration here.

The semantic property that corresponds to **A8**$_{IB}$ is:

$$\forall s, t, u((sI_i t \wedge sB_i u) \Rightarrow uI_i t),$$

while analogously **A8**$_{GB}$ corresponds to:

$$\forall s, t, u((sG_i t \wedge sB_i u) \Rightarrow G_i t)$$

Proof for A8$_{IB}$ We need to prove for all frames $F = (W, \{B_i : i \in \mathcal{A}\}, \{G_i : i \in \mathcal{A}\}, \{I_i : i \in \mathcal{A}\})$, that $F \models \neg\text{INT}(i, \varphi) \rightarrow \text{BEL}(i, \neg\text{INT}(i, \varphi))$ if and only if:

$$\forall s, t, u \in W((sI_i t \wedge sB_i u) \Rightarrow uI_i t)$$

For the easy direction from right to left, assume that:

$$\forall s, t, u \in W((sI_i t \wedge sB_i u) \Rightarrow uI_i t)$$

holds in a Kripke frame F. Now take any valuation Val on the set of worlds W and let $\mathcal{M}$ be the Kripke model arising from F by adding Val.

Now take any $s \in W$ with $\mathcal{M}, s \models \neg\text{INT}(i, \varphi)$, then there is a $t \in W$ with $sI_i t$ and $\mathcal{M}, t \not\models \varphi$. We will show that $\mathcal{M}, s \models \text{BEL}(i, \neg\text{INT}(i, \varphi))$. So take any $u \in W$ such that $sB_i u$. By the semantic property of the frame, we have $uI_i t$, so $\mathcal{M}, u \models \neg\text{INT}(i, \varphi)$ and indeed $\mathcal{M}, s \models \text{BEL}(i, \neg\text{INT}(i, \varphi))$. Therefore:

$$F \models \neg\text{INT}(i, \varphi) \rightarrow \text{BEL}(i, \neg\text{INT}(i, \varphi))$$

For the other direction, work by contraposition and suppose that the semantic property does not hold in a certain frame F. Then there are worlds s, t, u in the set of worlds W such that $sI_i t$ and $sB_i u$ but *not* $uI_i t$.

Now the valuation Val on F such that for all $v \in W$, $Val(p) = 1$ iff $uI_i v$ and let $\mathcal{M}$ be the Kripke model arising from F by adding Val. Then by definition $\mathcal{M}, t \not\models p$, so in turn $\mathcal{M}, s \models \neg\text{INT}(i, p)$. On the other hand, $\mathcal{M}, u \models \text{INT}(i, p)$, so $\mathcal{M}, s \not\models \text{BEL}(i, \neg\text{INT}(i, p))$. We may conclude that:

$$F \not\models \neg\text{INT}(i, p) \rightarrow \text{BEL}(i, \neg\text{INT}(i, p))$$

The proofs for **A8**$_{GB}$, **A7**$_{IB}$ and **A7**$_{GB}$ are similar.

3.3.1.3 Relating Intentions to Goals

We assume that every intention corresponds to a goal:

A9$_{IG}$ $\text{INT}(i, \varphi) \rightarrow \text{GOAL}(i, \varphi)$ (Intention implies goal)

This means that if an agent adopts a formula as an intention, it should have adopted that formula as a goal to achieve, which satisfies Bratman's notion that an agent's intentions form a specific, in fact by **A6**$_I$ consistent, subset of its goals (Bratman, 1987).

Rao and Georgeff (1991) adopt this axiom as *goal-intention compatibility* for their class of O-formulas. In our non-temporal context, the corresponding semantic property is as follows.

3.3.1.4   Fact

The semantic property corresponding to $\mathbf{A9}_{IG}$ is that $G_i \subseteq I_i$.

Proof for $\mathbf{A9}_{IG}$ We need to prove for all frames $F = (W, \{B_i : i \in \mathcal{A}\}, \{G_i : i \in \mathcal{A}\}, \{I_i : i \in \mathcal{A}\})$, that $F \models \text{INT}(i, \varphi) \to \text{GOAL}(i, \varphi)$ if and only if $G_i \subseteq I_i$. For the easy direction from right to left, assume that $G_i \subseteq I_i$ holds in a Kripke frame F. Now take any valuation Val on the set of worlds W and let M be the Kripke model arising from F by adding Val. Now take any $s \in W$ with $M, s \models \text{INT}(i, \varphi)$, but suppose, in order to derive a contradiction, that $M, s \not\models \text{GOAL}(i, \varphi)$. Then there is a $t \in W$ with sG_it and $M, t \not\models \varphi$. But because $G_i \subseteq I_i$ we have sI_at as well, contradicting the assumption $M, s \models \text{INT}(i, \varphi)$. Therefore, $F \models \text{INT}(i, \varphi) \to \text{GOAL}(i, \varphi)$.

For the other direction, work by contraposition and suppose that $G_i \subseteq I_i$ does *not* hold in a certain frame F. Then there are worlds s, t in the set of worlds W such that sG_it but *not* sI_it. Now define the valuation Val on F such that for all $v \in W$, $Val(p) = 1$ iff sI_iv, and let M be the Kripke model arising from F by adding Val. Then by definition, we have $M, s \models \text{INT}(i, p)$; but $M, t \not\models p$ because not sI_it, so $M, s \not\models \text{GOAL}(i, p)$. We may conclude that $F \not\models \text{INT}(i, p) \to \text{GOAL}(i, p)$.

Remark 3.1 The correspondences presented in the above facts can also be quickly and nicely proved using the technique of second-order quantifier elimination (Gabbay et al., 2008).[2]

Definition 3.4 *The full system for individual attitudes, including both the axioms for informational attitudes given in Section 2.4 and the axioms for motivational attitudes and interdependencies given above, will be called* TEAMLOG^{ind} *(the individual part of teamwork logic) in the sequel.*

As to the side-effect problem, note that TEAMLOG^{ind} fortunately does *not* prove that an agent intends all the consequences it *believes* its intentions to have, that is the believed side-effects of its intentions.

Thus, TEAMLOG$^{ind} \not\vdash \text{BEL}(i, \varphi \to \psi) \to (\text{INT}(i, \varphi) \to \text{INT}(i, \psi))$.

There is a weaker version that does hold, though, namely if $\models \varphi \to \psi$ (a quite strong assumption!), then by $\mathbf{R2}_I$, we have $\models \text{INT}(i, \varphi) \to \text{INT}(i, \psi)$. Therefore, agents intend all logical consequences of their intentions. This is similar to the logical omniscience problem for logics of knowledge and belief discussed in Section 2.7.1. For a discussion of the 'side-effect problem' for intentions, see Bratman (1987), Cohen and Levesque (1990) and Rao and Georgeff (1991).

3.4   Collective Intention Constitutes a Group

Collective intention, as a specific joint mental attitude, is the central topic addressed in teamwork. We agree with Levesque *et al.* (1990) that:

> Joint intention by a team does not consist merely of simultaneous and coordinated individual actions; to act together, a team must be aware of and care about the status of the group effort as a whole.

---

[2] These and similar correspondences can be automatically computed at http://www.fmi.uni-sofia.bg/fmi/logic/sqema/index.jsp.

In TEAMLOG, teams are created on the basis of *collective intentions*: a team is constituted as soon as a collective intention among the members is present and stays together as long as the collective intention persists. In this chapter we focus on defining several notions of collective intentions in Section 3.5 and Section 3.7, abstracting from the team formation process. We refer the interested reader to Chapter 8 and Castelfranchi *et al.* (1992), Dignum *et al.* (2001a, b), Jennings (1993) and Wooldridge and Jennings (1999).

In contrast to Cohen and Levesque (1990), we are interested in generic characteristics of intentions, resigning from classifying them further along different dimensions. The collective choice that is 'hidden' in group intention directly leads to a collective commitment. The essential characteristics of commitments follow the linguistic tradition that while intentions ultimately lead to actions, the immediate triggers of these actions are commitments. In fact, social (or bilateral) commitments are related to individual actions, while collective commitments are related to plan-based team actions.

As mentioned before, we agree with Bratman (1987) that in human practical reasoning, intentions are first class citizens, that are not reducible to beliefs and desires. They form a rather special consistent subset of an agent's goals, that it wants to focus on for the time being. This way they create a screen of admissibility for the agent's further, possibly long-term, deliberation. In this chapter we extend this view to the collective intention case. In TEAMLOG, collective intentions are not introduced as primitive modalities, with some restrictions on the semantic accessibility relations (as in, for example, Cavedon *et al.* (1997)). We do give necessary and sufficient conditions for collective motivational attitudes to be present, making teamwork easier to predict. Collective commitments are treated in Dunin-Kȩplicz and Verbrugge (1996, 1999) and most extensively in Chapter 4.

In the philosophical and MAS literature there is an ongoing discussion as to whether collective intentions may be reduced to individual ones plus common beliefs about them (see Castelfranchi (1995), Haddadi (1995) and Tuomela and Miller (1988)). Even though our definition in Section 3.5 seems to be reductive, it involves infinitely nested intentions and group epistemic operators, making them much deeper than a simple compound built out of individual intentions and common beliefs by propositional connectives only.

Despite the overall complexity of collective motivational notions, in our investigation we tried to find *minimal* conditions characterizing them, and not to weigh down the definitions with all possible aspects applicable in specific situations. Such elements as conventions, abilities, opportunities, power relations and social structure (see Singh (1997), Tuomela (1995) and Wooldridge and Jennings (1999) for a thorough discussion) certainly are important. Therefore, we leave open the possibility of extending TEAMLOG by additional properties. For example, abilities and opportunities are important in dialogues recognizing potential towards a specific goal. These dialogues will be discussed in Chapter 8 (see also Dignum *et al.* (2001b)). Power relations and social structure, on the other hand, are reflected in collective commitments (see Chapter 4).

3.5   Definitions of Mutual and Collective Intentions

In this book, we focus on strictly cooperative teams, where 'cooperative' is meant in a stronger sense than the homonymous concept in game theory. See Bratman (1999) and Tuomela (1995) for good philosophical discussions on strong types of cooperation and collaboration needed in teamwork. This essence of cooperation makes the concept of collective intention rather powerful.

So, what motivates a group of agents to combine their efforts to achieve a given goal φ? First, they all need to individually *intend* φ. This leads to the so-called *general intention* E-INT$_G(\varphi)$ in the group G (see Table 3.2 for the relevant formulas). In fact, this necessary condition is taken to fully characterize collective intention in Rao *et al.* (1992) (see Wooldridge and Jennings (1996) for a similar definition of collective goal). However, this is certainly not sufficient. Imagine that two agents want to reach the same goal but are in a competition, willing to achieve it exclusively. Therefore, to exclude cases of competition, all agents should *intend* all members to have the associated individual intention, as well as the intention that all members have the individual intention, and so on. This simply means that general intention should be iterated in order to express the reciprocity of this process: 'everyone intends that everyone intends that everyone intends that ... φ'. We will call the resulting attitude a *mutual intention* M-INT$_G(\varphi)$.

Table 3.2   Group formulas and their intended meanings.

E-INT$_G(\varphi)$	Every agent in group G has the individual intention to achieve φ
M-INT$_G(\varphi)$	Group G has the mutual intention to achieve φ
C-INT$_G(\varphi)$	Group G has the collective intention to achieve φ

Even though mutual intention creates the motivational core of group intention, it would't be enough, if the agents weren't aware about their mutual attitudes. Thus, group members need to be aware of their reciprocal intentions. As discussed in the previous chapter, there are many ways of defining group awareness. Paradigmatically, in teamwork it is expressed by common belief C-BEL$_G$(M-INT$_G(\varphi)$), which we choose for the time being. This way a loosely coupled group becomes a strictly cooperative team. Of course, team members remain autonomous in maintaining their other motivational attitudes and may even be in competition about other issues.

In order to formalize the above conditions, a *general intention* E-INT$_G(\varphi)$ (standing for 'everyone intends') is defined by the following axiom, corresponding to the semantic condition that $\mathcal{M}, s \models$ E-INT$_G(\varphi)$ iff for all $i \in G$, $\mathcal{M}, s \models$ INT(i, φ):

M1   E-INT$_G(\varphi) \leftrightarrow \bigwedge_{i \in G}$ INT(i, φ).

The mutual intention M-INT$_G(\varphi)$ is meant to be true if everyone in G intends φ, everyone in G intends that everyone in G intends φ, etc. As we do not have infinite formulas to express this, let E-INT$_G^1(\varphi)$ be an abbreviation for E-INT$_G(\varphi)$, and let E-INT$_G^{k+1}(\varphi)$ for $k > 1$ be an abbreviation of E-INT$_G$(E-INT$_G^k(\varphi)$). Thus we have $\mathcal{M}, s \models$ M-INT$_G(\varphi)$ iff $\mathcal{M}, s \models$ E-INT$_G^k(\varphi)$ for all $k \geq 1$.

Define world t to be G_I-*reachable* from world s iff $(s, t) \in (\bigcup_{i \in G} I_i)^+$, the transitive closure of the union of all individual accessibility relations. Formulated more informally, this means that there is a path of length ≥ 1 in the Kripke model from s to t along accessibility arrows I_i that are associated with members i of G.

Then the following property holds (see Section 2.4 and Fagin *et al.* (1995) for analogous properties for common belief and common knowledge, respectively):

$$\mathcal{M}, s \models \text{M-INT}_G(\varphi) \text{ iff } \mathcal{M}, t \models \varphi \text{ for all } t \text{ that are } G_I\text{-reachable from } s$$

Using this property, it can be shown that the following fixed-point axiom and rule can be soundly added to the union of KD_n and **M1**:

M2 $M\text{-INT}_G(\varphi) \leftrightarrow E\text{-INT}_G(\varphi \wedge M\text{-INT}_G(\varphi))$

RM1   From $\varphi \rightarrow E\text{-INT}_G(\psi \wedge \varphi)$ infer $\varphi \rightarrow M\text{-INT}_G(\psi)$ (Induction Rule)

The resulting system is called TEAMLOG^{mint} (the part of teamwork logic for mutual intentions) and is sound and complete with respect to Kripke models where all n accessibility relations for intentions are serial. The completeness proof will be given in Section 3.8. Now we will show the soundness of Rule **RM1** with respect to the given semantics. (The other axioms and rules of TEAMLOG^{mint} are more intuitive, so we leave their soundness to the reader.)

Assume that $\models \varphi \rightarrow E\text{-INT}_G(\psi \wedge \varphi)$, meaning that $\varphi \rightarrow E\text{-INT}_G(\psi \wedge \varphi)$ holds in all worlds of all Kripke models. We need to show that $\models \varphi \rightarrow M\text{-INT}_G(\psi)$. So take any Kripke model $\mathcal{M} = (W, \{B_i : i \in \mathcal{A}\}, \{G_i : i \in \mathcal{A}\}, \{I_i : i \in \mathcal{A}\}, Val)$ with $\mathcal{A} = \{1, \ldots, n\}$, and any world $s \in W$ with $\mathcal{M}, s \models \varphi$. Now assume that t is G_I-reachable from s in k steps along the path $w_0, \ldots w_k$ with $w_0 = s$ and $w_k = t$, by $k \geq 1$ relations of the form I_j ($j \in \{1, \ldots, n\}$). We need to show that $\mathcal{M}, t \models \psi$, for which we can show step by step that $\psi \wedge \varphi$ holds in all worlds w_i, $i \geq 1$, on the path from s to t. For the first step, for example $s I_j w_1$, we can use the fact that $\mathcal{M}, s \models \varphi \rightarrow E\text{-INT}_G(\psi \wedge \varphi)$, and thus $\mathcal{M}, s \models E\text{-INT}_G(\psi \wedge \varphi)$, to conclude that $\mathcal{M}, w_1 \models \psi \wedge \varphi$. Repeating this reasoning on the path to t, we conclude that for all i with $1 \leq i \leq k$, $\mathcal{M}, w_i \models \psi \wedge \varphi$, in particular $\mathcal{M}, t \models \psi$. We conclude that $\mathcal{M}, s \models \varphi \rightarrow M\text{-INT}_G(\psi)$, as desired.

Finally, the collective intention is defined by the following axiom:

M3 $C\text{-INT}_G(\varphi) \leftrightarrow M\text{-INT}_G(\varphi) \wedge C\text{-BEL}_G(M\text{-INT}_G(\varphi))$

The definition would be even stronger if common knowledge were applied in **M3**. However, because common knowledge is almost impossible to establish in multi-agent systems due to the unreliability of communication media (see Chapter 2), we do not pursue this strengthening further.

Definition 3.5   *The resulting system, which we call* TEAMLOG, *is the union of* TEAMLOG^{mint} *(for mutual intentions),* $KD45_n^C$ *(for common beliefs) and axiom* **M3**.

3.5.1   Some Examples

Let us give an informal example of the establishment of a collective intention. Two violinists, a and b, have studied together and have toyed with the idea of giving a concert together someday. Later this has become more concrete: they both intend to perform the two solo parts of the Bach Double Concerto, expressed in $INT(a, \varphi)$ and $INT(b, \varphi)$, where φ stands for 'a and b perform the solo parts of the Bach Double Concerto'. After communicating with each other, they start practising together. Clearly, a mutual intention $M\text{-INT}_{\{a,b\}}(\varphi)$ as defined in **M2** is now in place, involving nested intentions like $INT(a, INT(b, INT(a, \varphi)))$ and so on. The communication established a common belief $C\text{-BEL}_G(M\text{-INT}_G(\varphi))$ about their mutual intention with $G = \{a, b\}$, according to **M3**. As sometimes happens in life, when people are ready, an opportunity appears: Carnegie Hall plans a concert for Christmas Eve, including the Bach Double Concerto. Now they refine their collective intention to a more concrete $C\text{-}INT_G(\psi)$ (where ψ stands for 'a and b perform the solo parts of the Bach Double Concerto at the Christmas Eve concert in Carnegie Hall'). Luckily, our two violinists are chosen from among a list of candidates to be the soloists, and both sign the appropriate contract. Because they do this together, common knowledge, not merely common belief of their mutual intention is present:

$$M\text{-}INT_G(\psi) \wedge C\text{-}KNOW_G(M\text{-}INT_G(\psi))$$

One important difference between common knowledge and common belief is that common knowledge can be justified if needed and a commonly signed contract provides a perfect basis for this. Clearly, the two violinists have developed a very strong variant of collective intention due to their common knowledge of the mutual intention.

3.5.2 Collective Intentions Allow Collective Introspection

The following lemma about positive introspection for collective intentions follows easily from the definition of collective intention, using Lemma 2.1, as we will show below.

Lemma 3.1 *Let φ be a formula and $G \subseteq \{1, \ldots, n\}$. Then the principle of* collective *positive introspection of collective intentions holds:*

$$C\text{-}INT_G(\varphi) \rightarrow C\text{-}BEL_G(C\text{-}INT_G(\varphi))$$

Proof We give a syntactic proof sketch. By **M3**, we have:

$$\text{TEAMLOG} \vdash C\text{-}INT_G(\varphi) \rightarrow C\text{-}BEL_G(M\text{-}INT_G(\varphi))$$

Then, by the second part of Lemma 2.1 about positive introspection for common beliefs, we have:

$$\text{TEAMLOG} \vdash C\text{-}BEL_G(M\text{-}INT_G(\varphi)) \rightarrow C\text{-}BEL_G C\text{-}BEL_G((M\text{-}INT_G(\varphi))$$

Combining these two, we get:

$$\text{TEAMLOG} \vdash C\text{-}INT_G(\varphi) \rightarrow C\text{-}BEL_G(M\text{-}INT_G(\varphi)) \wedge$$
$$C\text{-}BEL_G(C\text{-}BEL_G(M\text{-}INT_G(\varphi)))$$

So by the first part of Lemma 2.1 about distribution of common beliefs over conjunctions, we have:

$$\text{TEAMLOG} \vdash C\text{-}INT_G(\varphi) \rightarrow C\text{-}BEL_G(M\text{-}INT_G(\varphi) \wedge C\text{-}BEL_G(M\text{-}INT_G(\varphi)))$$

which leads, by **M3**, to the desired:

$$\text{TEAMLOG} \vdash C\text{-}INT_G(\varphi) \rightarrow C\text{-}BEL_G(C\text{-}INT_G(\varphi))$$

3.6 Collective Intention as an Infinitary Concept

Due to the fixed-point axiom **M2** and the induction rule **RM1**, the finite theory TEAMLOG can capture the potentially infinitary concept of mutual intention. Then again, the concept of collective intention as defined in axiom **M3** is based on the potentially infinitary concept of common belief. How to approach the seeming tensions between an infinitary concept and a finite logical theory?

3.6.1 Mutual Intention is Created in a Finite Number of Steps

Even though M-INT$_G(\varphi)$ is an infinite concept, mutual intentions may be established in practice in a finite number of steps. Axiom **M2** makes evident that it suffices for a mutual intention that all potential team members intend φ and that they accept the individual intention towards a mutual intention to achieve φ, in order to foster cooperation from the start.

Formally, for every $i \in G$, only INT$(i, \varphi \wedge$ M-INT$_G(\varphi))$ needs to be established. This implies by axiom **M1** that E-INT$_G(\varphi \wedge$ M-INT$_G(\varphi))$, which in turn implies by axiom **M2** that M-INT$_G(\varphi)$ holds. Notice that in contrast to the logically similar common beliefs, the creation of mutual intentions does not necessarily depend on the communication medium: simply all individual agents need to appropriately change their minds.

The tricky part of collective intentions that does depend on communication, is the awareness-part, namely C-BEL$_G($M-INT$_G(\varphi))$. Thus, standard collective intentions defined by **M3** are appropriate to model those situations in which communication, in particular announcements, work, especially if one initiator establishes the team. In Chapter 9 we show in detail how team formation in an ideal case may be actually realized through complex dialogues.

The standard definition is also applicable in situations foreseen during the design phase. For example, emergency scenarios with classified fixed protocols and the roles predefined accordingly, like a yacht on the sea. Thus, in specific circumstances, team members may know in advance their roles in predefined scenarios, and have individual intentions to fulfill them, as well as to achieve the main goal. They also intend others to fulfill their intentions, etc: E-INT$_G(\varphi \wedge$ M-INT$_G(\varphi))$. Therefore, the mutual intention M-INT$_G(\varphi)$ is present immediately. This factor is essential when people or even precious goods or equipment are in danger!

It is interesting to investigate whether one could do in general with only one or two levels of general intention E-INT$_G$ to cover teamwork. Indeed, such proposals have been made in the MAS literature; let us discuss two of them.

3.6.2 Comparison with the One-Level Definition

In order to verify the correctness of Definition **M3**, one needs to check whether inappropriate cases are not unintentionally covered. Thus, collective intention shouldn't cover situations where real teamwork is out of the question. Bratman (1999, Chapter 5) characterizes shared cooperative activity. Moreover, he gives some exemplar situations where, even though agents share some attitudes, their cooperative activity is excluded. Fortunately, our definition excludes these cases, as well. For example:

> Suppose that you and I each intend that we go to New York together, and this is known to both of us. However, I intend that we go together as a result of my kidnapping you and forcing you to join me. The expression of my intention, we might say, is the Mafia sense of 'We're going to New York together'.[3] While I intend that we go to New York together, my intentions are clearly not cooperative in spirit (Bratman, 1999).

---

[3] One may criticize Bratman's formulation of the example. In logic, if the two agents attach a different meaning to 'we go to New York together', then the two meanings should be expressed differently; see for example, Montague (1973). In the current set-up, b intends something like 'a and b go to New York by b forcing b', while a intends

Now, take $\varphi =$ 'a and b go to New York' with a for 'you' and b for 'me' and let G stand for $\{a, b\}$. In the Mafia situation sketched above, the two agents do have a general intention E-INT$_G(\varphi)$ and possibly also a common belief C-BEL$_G$(E-INT$_G(\varphi)$) holds, but neither M-INT$_G(\varphi)$, nor C-INT$_G(\varphi)$ is present. Specifically, it seems unlikely that INT(b, INT(a, φ)) holds for the Mafioso.

Note that Rao, Georgeff and Sonenberg's definition of a *joint intention* among G to achieve φ is defined as E-INT$_G(\varphi) \wedge$ C-BEL$_G$(E-INT$_G(\varphi)$) (translated to our notation); thus it erroneously ascribes a joint intention to go to New York among the agents in the example (Rao *et al.*, 1992). Incidentally, a similar one-level definition of mutual goals was also given by Wooldridge and Jennings (1996). These definitions do not even exclude cases of individual competition in the course of potential teamwork.

3.6.3   Comparison with the Two-Level Definition

In previous work, we gave a somewhat weaker definition of collective intention than the one in Section 3.5 (see Dunin-Kęplicz and Verbrugge (1996, 1999)). While attempting to build a possibly simple definition, its expressive power turned out to be too limited. The definition consisted of two levels of reciprocal intentions in a team and a common belief about this. Thus, fortunately, it did not erroneously assign a collective intention to sets whose members are in individual competition or in coercive situation, as in the one-level definition discussed above. Our obsolete two-level definition is the following:

$$C\text{-INT}_G^{old}(\varphi) \leftrightarrow E\text{-INT}_G(\varphi) \wedge C\text{-BEL}_G(E\text{-INT}_G(\varphi))$$

$$\wedge \, E\text{-INT}_G(E\text{-INT}_G(\varphi)) \wedge C\text{-BEL}_G(E\text{-INT}_G(E\text{-INT}_G(\varphi)))$$

In words, a group would have a collective intention if everyone intends to achieve the goal φ and there is a common belief about this general intention (level 1) and in addition, everyone intends there to be the general intention towards φ together with a common belief about this (level 2).

Unfortunately, however, the above definition did not preclude competition among more-person coalitions. Consider the following example. Three world-famous violinists a, b and c are candidates to be one of the two lead players needed to play the Bach Double Concerto, for a performance in Carnegie Hall on Christmas Eve. They are asked to decide among themselves who will be the two soloists. Imagine the situation where all three of them want to be one of the 'chosen two', and they also want both other players to want this – as long as it is with them, not with the third player. For example, a is against a coalition between b and c to play the violin concerto together as soloists.

Thus, for $\varphi(i) =$ 'there will be a great performance of the Bach Double Concerto in Carnegie Hall on Christmas Eve including soloist i', we have two levels for reciprocal intention among pairs from $\{a, b, c\}$ (for example, INT(a, INT($b, \varphi(b)$))), and even M-INT$_{\{a,b\}}\varphi(b)$). But we do not have a third one: a does not intend that b intends c to intend $\varphi(c)$ (so there is no M-INT$_{\{a,b,c\}}\varphi(c)$). Thus one would hardly say that a collective intention among them is in place: they are not a team, but rather three competing coalitions of two violinists each.

---

something like 'a and b go to New York by mutual consent'. In the formalization below the quote, we are charitable by translating 'we go to New York together' to the more neutral 'a and b go to New York'.

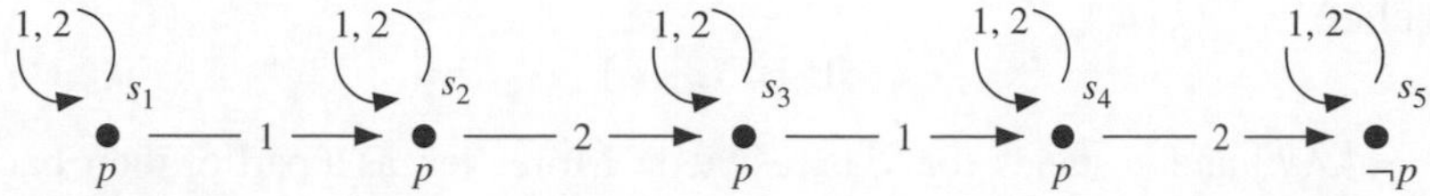

Figure 3.1 KD_2 model with accessibility relations I_i (represented by arrows labeled with the respective agent names). The following hold for $G = \{1, 2\}$: $\mathcal{M}, s_1 \models \text{E-INT}_G^1(p)$, $\mathcal{M}, s_1 \models \text{E-INT}_G^2(p)$ and $\mathcal{M}, s_1 \models \text{E-INT}_G^3(p)$; however, $\mathcal{M}, s_1 \not\models \text{E-INT}_G^4(p)$ and therefore certainly $\mathcal{M}, s_1 \not\models \text{M-INT}_G(p)$.

3.6.4 Can the Infinitary Concept be Replaced by a Finite Approximation?

If we adapt the definition above to make it consist of three levels of intention instead of two, the troublesome example of the two-agent coalitions would be solved. However, one may invent similar (admittedly artificial) examples for any k, using coalitions of k people from among a base set of at least $k + 1$ agents. Thus, the infinitary mutual intention of Section 3.5 was derived to avoid all such counterexamples.

In many practical cases, one can manage with a fixed stack of general intentions. Also theoretically, it has been proved that in Kripke models with a fixed bound of at most k worlds, $\text{E-KNOW}_G^k(\varphi)$ is equivalent to $\text{C-KNOW}_G(\varphi)$ (see Exercise 2.2.10 in Meyer and van der Hoek (1995)). The same reasoning holds for mutual intentions: on models of at most k worlds, $\text{E-INT}_G^k(\varphi)$ is equivalent to $\text{M-INT}_G(\varphi)$. The level needed is independent of the number of agents involved. However, it turns out that $\text{M-INT}_G(\varphi)$ cannot be reduced to a fixed level of general intention: for any given number of agents and any k, we can find a Kripke model (of size larger than k) such that $\text{M-INT}_G(\varphi)$ is *not* equivalent to $\text{E-INT}_G^k(\varphi)$. Figure 3.1 shows a counterexample for two agents and $k = 4$.

3.7 Alternative Definitions

Even though the notion of collective intention is idealized, it can be adjusted to the circumstances of the application. Let us give some examples.

3.7.1 Rescue Situations

In some situations, especially time-critical ones, we will argue that teamwork may tentatively start even if the standard collective intention (Definition **M3** in Section 3.5) has not yet been established. Actually, it may happen that a mutual intention is naturally established, in contrast to a common belief about this. In order to initiate teamwork in such situations, a modified notion of group intention can be of use.

Consider, for example, a situation in which a person c has disappeared under the ice and two potential helpers a and b are in the neighbourhood and run towards the person in danger. They do not know each other, and there is no clear initiator among them. Assume further that, at this point in time, communication among them is not possible, due to strong wind and the distance between them. On the other hand, visual perception is possible in a limited way: they can see each other move but cannot distinguish facial expressions. Both intend to help and thus $INT(a, \varphi)$ and $INT(b, \varphi)$ hold, as well as a general belief:

$$\text{E-BEL}_G(\varphi)$$

where $G = \{a, b\}$ and φ stands for 'c is rescued'. Moreover, as a part of their background knowledge, they know that in general two persons are needed for a successful rescue, in fact, it is a common belief. Also, it seems to be justified to assume that the other person knows this fact as well, in fact, it is a common belief. Thus:

$$\text{C-BEL}_G(\psi)$$

holds, where ψ stands for 'at least two persons are needed to rescue someone disappearing under the ice'.

As there are no other potential helpers around, a and b believe that they need to act together. Thus, we may naturally expect that a mutual intention is in place:

$$\text{M-INT}_G(\varphi)$$

Both agents may even form an individual belief about the mutual intention, so at this point there may be:

$$\text{M-INT}_G(\varphi) \wedge \text{E-BEL}_G(\text{M-INT}_G(\varphi))$$

However, communication being limited, the common belief about the mutual intention $\text{C-BEL}_G(\text{M-INT}_G(\varphi))$ cannot be established; for this reason, the standard collective intention $\text{C-INT}_G(\varphi)$ does not hold. In the rescue situation, such a common belief enables coordination needed for mouth-on-mouth breathing and heart massage. As time is critical, even if communication is severely restricted at present, the two agents still try to establish a team together, and both *intend* that the common belief about the mutual intention be established to make real teamwork possible. Thus, it is justified to base a goal-directed activity on a somewhat revised notion of mutual intention.

Therefore, we define a notion that is somewhat stronger than the mutual intention defined in axiom **M2** but yet does not constitute a proper collective intention. In axiom **M2′** below, even though a common belief about the mutual intention has actually not yet been established, all members of the group intend it to be in place.

Thus, the alternative mutual intention $\text{M-INT}'_G(\varphi)$ is meant to be true if everyone in G intends φ, everyone in G intends that everyone in G intends φ, etc., as in $\text{M-INT}_G(\varphi)$; moreover, everyone intends that there be a common belief in the group of the mutual intention: $\text{E-INT}_G(\text{C-BEL}_G(\text{M-INT}_G(\varphi)))$. This is reflected by the following axiom, which can be soundly added to TEAMLOG (for standard collective intentions):

M2′ $\text{M-INT}'_G(\varphi) \leftrightarrow (\text{M-INT}_G(\varphi) \wedge \text{E-INT}_G(\text{C-INT}_G(\varphi)))$

The resulting system is called $\text{KD}_n^{\text{M-INT}'_G}$, and it can easily be seen to be sound with respect to Kripke models where all n accessibility relations for all I_i and B_i ($i \in \{1, \ldots, n\}$) are serial, while those for the B_i are additionally transitive and Euclidean.

The notion of $\text{M-INT}'_G$ is appropriate for unstable situations in which communication is hardly possible, while team action is essential. From this perspective, $\text{M-INT}'_G$ may be called a 'pre-collective intention', useful as a precursor for a collective intention to be established.

Note that $\text{M-INT}'_G$ includes intentions about awareness, whereas in the original definition of C-INT_G awareness exists, whether or not intended. Therefore $\text{M-INT}'_G$ is stronger than M-INT$_G$, but it is not comparable to C-INT$_G$. Formally, the implications may be represented as follows:

$$\vdash \text{C-INT}_G(\varphi) \to \text{M-INT}_G(\varphi)$$

$$\vdash \text{M-INT}'_G(\varphi) \to \text{M-INT}_G(\varphi)$$

$$\nvdash \text{M-INT}'_G(\varphi) \to \text{C-INT}_G(\varphi)$$

$$\nvdash \text{C-INT}_G(\varphi) \to \text{M-INT}'_G(\varphi)$$

The proofs are straightforward and are left to the reader.

3.7.2   Tuning Group Intentions to the Environment

In the standard definition of collective intention, it was stipulated that in order to turn a mutual intention into a collective one, the team needs to have a common belief about it. However, we have just seen that in some circumstances the team has to do with a weaker kind of awareness. In other cases it can even create a stronger kind of awareness, such as common knowledge in the example about the contract between the violinists and Carnegie Hall.

In order to make the definition of collective intention more flexible, and thus to allow the system developer to tune his/her system to the environment at hand, we restate the definition as a scheme:

$$\textbf{M3}^{\text{schema}} \quad \text{C-INT}_G(\varphi) \leftrightarrow \text{M-INT}_G(\varphi) \wedge \textit{awareness}_G(\text{M-INT}_G(\varphi))$$

Instantiating the above schema corresponds metaphorically to tuning dials on a sound system. In this case, the $\textit{awareness}_G$-dials can be tuned from $\emptyset$ (no awareness at all), through individual beliefs INT for $i \in \{1, \ldots, n\}$ and different degrees of general beliefs E-BEL$_G^k$ ($k \geq 1$), to common belief C-BEL$_G$. This can analogously be done for degrees of knowledge.

The degree of awareness in $\textbf{M3}^{\text{schema}}$ clearly depends on the circumstances and varies from just recognizing the situation by perception when communication is difficult or impossible, through confirming what situation we deal with (then agents' predefined roles are clear), to more complex cases when, for example, some agents or roles are missing, so that more communication is needed.

An example of a collective intention where $\textit{awareness}_G$ is instantiated as E-BEL$_G$ occurs in the usual e-mail agreements where one person writes to another: 'Let us go to that movie *The Conclave* tonight' and the other replies 'OK', where both messages happen to arrive and no further acknowledgments are exchanged. After the interchange both parties believe in their mutual intention to go to that movie, so M-INT$_G(\varphi) \wedge$ E-BEL$_G$(M-INT$_G(\varphi)$) is achieved, but (because e-mail communication may be faulty) there is no common belief about their mutual intention.

3.8   The Logic of Mutual Intention TEAMLOG^{mint} is Complete

In this section, a completeness proof is given for TEAMLOG^{mint}, the logic of mutual intentions in the standard case (see Subsection 3.5). Soundness of TEAMLOG^{mint} with respect to serial models is easy to check. The most interesting part of the soundness proof, namely for induction rule **RM1**, was given in Section 3.5.

The completeness of TEAMLOG^{mint} enables the designer of a multi-agent system to test the validity of various properties concerned with collective intentions, by checking models instead of constructing axiomatic proofs. In addition, the completeness proof gives an upper bound on the complexity of reasoning about the satisfiability of such properties: by a 'small model theorem' for an analogous system, the problem has been shown to be in EXPTIME (see Fagin *et al.*, 1995). In Chapter 9, we show that the problem is also EXPTIME-hard, so it is EXPTIME-complete. The method of the completeness proof is one used often in modal logic when proving completeness with respect to *finite* models, for example when one shows decidability of a system. The proof is inspired by the one for the logic of common knowledge in Fagin *et al.* (1995), which is in turn inspired by Parikh's and Kozen's completeness proof for propositional dynamic logic (Kozen and Parikh, 1981).[4] In fact, the main difference consists in adapting their proof to our slightly different choice of axioms and filling in some steps that were left to the reader in Fagin *et al.* (1995).

We have to prove that, supposing that TEAMLOG^{mint} $\nvdash \varphi$, there is a serial model M and a $w \in M$ such that $M, w \nvDash \varphi$. We fix the number m of available agents throughout this section and suppress the subscript m for TEAMLOG$^{mint}_m$. There will be four steps:

1. A finite set of formulas Φ, the *closure* of φ, will be constructed that contains φ and all its subformulas, plus certain other formulas that are needed in Step 4 below to show that an appropriate valuation falsifying φ at a certain world can be defined. The set Φ is also closed under single negations.
2. A Finite Lindenbaum lemma will be proved: a consistent finite set of sentences from Φ can always be extended to a finite set that is maximally consistent in Φ.
3. These finitely many maximally consistent sets will correspond to the states in the Kripke countermodel against φ and appropriate accessibility relations and a valuation will be defined on these states.
4. It will be shown, using induction on all formulas in Φ, that the model constructed in Step 3 indeed contains a world in which φ is false. This is the most complex step in the proof.

Below, the closure of a sentence φ is defined. One can view it as the set of formulas that are *relevant* for making a countermodel against φ. It is similar to the well-known Fischer--Ladner closure (Fischer and Ladner, 1979).

Definition 3.6 *The closure of φ with respect to TEAMLOG^{mint} is the minimal set Φ of TEAMLOG^{mint}-formulas such that for all $G \subseteq \{1, \ldots, m\}$ the following hold:*

1. $\varphi \in \Phi$.
2. *If $\psi \in \Phi$ and χ is a subformula of ψ, then $\chi \in \Phi$.*
3. *If $\psi \in \Phi$ and ψ itself is not a negation, then $\neg\psi \in \Phi$.*
4. *If M-INT$_G(\psi) \in \Phi$ then E-INT$_G(\psi \wedge$ M-INT$_G(\psi)) \in \Phi$.*

---

[4] To complete the historical roots, a complete set of axioms for PDL was first proposed by Segerberg (1977). A completeness proof for another axiomatization appeared in Fischer and Ladner (1979). Completeness for Segerberg's axiomatization was first proved independently by Parikh (1978) and Gabbay (1977).

5. *If* E-INT$_G(\psi) \in \Phi$ *then* INT$(i, \psi) \in \Phi$ *for all* $i \in G$.
6. $\neg$INT$(i, \perp) \in \Phi$ *for all* $i \leq m$.

It is straightforward to prove that for every formula φ, the closure Φ of φ with respect to TEAMLOG^{mint} is a *finite* set of formulas.

This finishes step 1 of the completeness proof. The next definition leads up to the Finite Lindenbaum Lemma, Step 2 of the proof.

Definition 3.7 *A finite set of formulas* Γ *such that* $\Gamma \subseteq \Phi$ *is maximally* TEAMLOG^{mint}-*consistent in* Φ *if and only if*:

1. Γ *is* TEAMLOG^{mint}-*consistent, that is* TEAMLOG^{mint} $\not\vdash \neg(\bigwedge_{\psi \in \Gamma} \psi)$.
2. *There is no* $\Gamma' \subseteq \Phi$ *such that* $\Gamma \subset \Gamma'$ *and* Γ' *is still* TEAMLOG^{mint}-*consistent*.

Lemma 3.2 (Finite Lindenbaum Lemma) *Let* Φ *be the closure of* φ *with respect to* TEAMLOG^{mint}. *If* $\Gamma \subseteq \Phi$ *is* TEAMLOG^{mint}-*consistent, then there is a set* $\Gamma' \supseteq \Gamma$ *which is maximally* TEAMLOG^{mint}-*consistent in* Φ.

Proof By standard techniques of modal logic: enumerating all formulas in Φ and subsequently adding a formula or its negation depending on whether TEAMLOG^{mint}-consistency is preserved or not.

Now we are ready to take Step 3, namely to define the model that will turn out to contain a world where $\neg \varphi$ holds.

Definition 3.8 *Let* $M_\varphi = \langle W_\varphi, \{I_1, \ldots, I_m\}, Val \rangle$ *be a Kripke model defined as follows*:

- *As domain of states, one state* s_Γ *is defined for each maximally* TEAMLOG^{mint}-*consistent* $\Gamma \subseteq \Phi$. *Note that, because* Φ *is finite, there are only finitely many states. Formally, we define*:

$$\text{CON}_\Phi = \{\Gamma \mid \Gamma \text{ is maximally TEAMLOG}^{mint}\text{-consistent in } \Phi\} \text{ and}$$

$$W_\varphi = \{s_\Gamma \mid \Gamma \in \text{CON}_\Phi\}$$

- *To make a truth assignment* Val, *we want to conform to the propositional atoms that are contained in the maximally consistent sets corresponding to each world. Thus, we define* Val$(s_\Gamma)(p) = 1$ *if and only if* $p \in \Gamma$. *Note that this makes all propositional atoms that do not occur in* φ *false in every world of the model*.
- *The relations* I_i *are defined as follows*:

$$I_i = \{(s_\Gamma, s_\Delta) \mid \psi \in \Delta \text{ for all } \psi \text{ such that INT}(i, \psi) \in \Gamma\}$$

It will turn out that using this definition, we not only have $M_\varphi, s_\Gamma \models p$ iff $p \in \Gamma$ for propositional atoms p, but such an equivalence holds for all relevant formulas. This is proved in the Finite Truth Lemma, the main result of Step 4.

In order to prove the Finite Truth Lemma, we need to prove some essential properties of maximally TEAMLOG^{mint}-consistent sets in Φ, namely the Consequence Lemma and the Finite Valuation Lemma.

Lemma 3.3 (Consequence Lemma) *If* $\Gamma \in \text{CON}_\Phi$, *and moreover* $\psi_1, \ldots, \psi_n \in \Gamma$, $\chi \in \Phi$ *and* $\text{TEAMLOG}^{mint} \vdash \psi_1 \rightarrow (\psi_2 \rightarrow (\ldots (\psi_n \rightarrow \chi) \ldots))$, *then* $\chi \in \Gamma$.

Proof The proof is straightforward, by standard reasoning about maximal consistent sets (Meyer and van der Hoek, 1995).

Lemma 3.4 (Finite Valuation Lemma) *If* Γ *is* TEAMLOG^{mint}-*consistent in some closure* Φ, *then for all* ψ, χ *it holds that*:

1. *If* $\neg \psi \in \Phi$, *then* $\neg \psi \in \Gamma$ *iff* $\psi \notin \Gamma$.
2. *If* $\psi \wedge \chi \in \Phi$, *then* $\psi \wedge \chi \in \Gamma$ *iff* $\psi \in \Gamma$ *and* $\chi \in \Gamma$.
3. *If* $\text{INT}(i, \psi) \in \Phi$, *then* $\text{INT}(i, \psi) \in \Gamma$ *iff* $\psi \in \Delta$ *for all* Δ *with* $(s_\Gamma, s_\Delta) \in I_i$.
4. *If* $\text{E-INT}_G(\psi) \in \Phi$, *then* $\text{E-INT}_G(\psi) \in \Gamma$ *iff* $\psi \in \Delta$ *for all* Δ *and all* $i \in G$ *such that* $(s_\Gamma, s_\Delta) \in I_i$.
5. *If* $\text{M-INT}_G(\psi) \in \Phi$, *then* $\text{M-INT}_G(\psi) \in \Gamma$ *iff* $\psi \in \Delta$ *for all* Δ *that are* G_I-*reachable from* s_Γ.

Proof Items 1 and 2 are proved by standard modal logic techniques (Blackburn *et al.*, 2002; Meyer and van der Hoek, 1995). We will now prove 3, 4 and 5.
3: **the INT-case** Suppose $\text{INT}(i, \psi) \in \Phi$.

$\Rightarrow$ Assume $\text{INT}(i, \psi) \in \Gamma$, and assume that $(s_\Gamma, s_\Delta) \in I_i$. Then by definition of I_i, we immediately have $\psi \in \Delta$, as desired.

$\Leftarrow$ Suppose, by contraposition, that $\text{INT}(i, \psi) \notin \Gamma$. We need to show that there is a Δ such that $(s_\Gamma, s_\Delta) \in I_i$ and $\psi \notin \Delta$. It suffices to show the following **Claim**:

> **Claim:** The set of formulas $\Delta' = \{\chi \mid \text{INT}(i, \chi) \in \Gamma\} \cup \{\neg \psi\}$ is TEAMLOG^{mint}-consistent.

For if the claim is true, then by the Finite Lindenbaum Lemma there exists a maximally TEAMLOG^{mint}-consistent $\Delta \supseteq \Delta'$ in Φ. By the definitions of Δ' and I_i we have $(s_\Gamma, s_\Delta) \in I_i$, and by 1, we have $\psi \notin \Delta$, as desired. So let us prove the claim. In order to derive a contradiction, suppose Δ' is *not* TEAMLOG^{mint}-consistent. Because Δ' is finite, we may suppose that $\{\chi \mid \text{INT}(i, \chi) \in \Gamma\} = \{\chi_1, \ldots, \chi_n\}$. Then by the definition of inconsistency:

$$\text{TEAMLOG}^{mint} \vdash \neg(\chi_1 \wedge \ldots \wedge \chi_n \wedge \neg \psi)$$

By propositional reasoning, we get:

$$\text{TEAMLOG}^{mint} \vdash \chi_1 \rightarrow (\chi_2 \rightarrow (\ldots (\chi_n \rightarrow \psi) \ldots))$$

Then by necessitation (**R2$_I$**) plus a number of applications of (**A2$_I$**) and more propositional reasoning, we derive:

$$\text{TEAMLOG}^{mint} \vdash \text{INT}(i, \chi_1) \rightarrow (\text{INT}(i, \chi_2) \rightarrow (\ldots (\text{INT}(i, \chi_n) \rightarrow \text{INT}(i, \psi)) \ldots))$$

However, we know that $\text{INT}(i, \chi_1), \ldots, \text{INT}(i, \chi_n) \in \Gamma$ and $\text{INT}(i, \psi) \in \Phi$, so by the Consequence Lemma, $\text{INT}(i, \psi) \in \Gamma$, contradicting our starting assumption.

4: **the E-INT$_G$-case** Assume that E-INT$_G(\psi) \in \Phi$; then by the construction of Φ also INT$(i, \psi) \in \Phi$ for all $i \in G$.

$\Rightarrow$ Assume that E-INT$_G(\psi) \in \Gamma$. Axiom **M1** and some easy propositional reasoning gives us $T\textsc{eam}L\textsc{og}^{\text{mint}} \vdash$ E-INT$_G(\psi) \rightarrow$ INT(i, ψ) for all $i \in G$. Because INT$(i, \psi) \in \Phi$ we can use the Consequence Lemma and derive that INT$(i, \psi) \in \Gamma$ for all $i \in G$. Thus, by the $\Rightarrow$-step of the INT-case, we have $\psi \in \Delta$ for all Δ and all $i \in G$ such that $(s_\Gamma, s_\Delta) \in I_i$, as desired.

$\Leftarrow$ The proof is very similar to the $\Rightarrow$-step, this time using **M1** and the $\Leftarrow$-step of the INT-case.

5: **the M-INT$_G$-case** Let $s_\Gamma \longrightarrow^k s_\Delta$ stand for 's_Δ is G_I-reachable from s_Γ in k steps'. Assume that M-INT$_G(\psi) \in \Phi$; then by the construction of Φ also E-INT$_G(\psi \wedge$ M-INT$_G(\psi)) \in \Phi$, as well as its subformulas.

$\Rightarrow$ Assume that M-INT$_G(\psi) \in \Gamma$. We will prove by induction that for all $k \geq 1$ and all Δ, if $s_\Gamma \longrightarrow^k s_\Delta$, then ψ, M-INT$_G(\psi) \in \Delta$. (Note that this is stronger than what is actually needed for the $\Rightarrow$-step; such a loaded induction hypothesis makes the proof easier.)
$k = 1$ Assume that $s_\Gamma \longrightarrow^1 s_\Delta$; this means that $\Gamma I_i \Delta$ for some $i \in G$. By axiom **M2** we have $T\textsc{eam}L\textsc{og}^{\text{mint}} \vdash$ M-INT$_G(\psi) \rightarrow$ E-INT$_G(\psi \wedge$ M INT$_G(\psi))$.
So because M-INT$_G(\psi) \in \Gamma$ and E-INT$_G(\psi \wedge$ M-INT$_G(\psi)) \in \Phi$, the Consequence Lemma implies that E-INT$_G(\psi \wedge$ M-INT$_G(\psi)) \in \Gamma$. But then, by combining 4, the $\Rightarrow$-side of 3, and 2, we conclude that ψ, M-INT$_G(\psi) \in \Delta$, as desired.
$k = n + 1$ Assume that $s_\Gamma \longrightarrow^{n+1} s_\Delta$ for some $n \geq 1$, then there is a Δ' such that $s_\Gamma \longrightarrow^n s_{\Delta'}$ and $s'_\Delta \longrightarrow^1 s_\Delta$. By the induction hypothesis, we have ψ, M-INT$_G(\psi) \in \Delta'$. Now, just as in the base case $k = 1$, one can prove that the formulas ψ, M-INT$_G(\psi)$ are transferred from Δ' to the direct successor Δ.

$\Leftarrow$ This time we work directly, not by contraposition. So assume that $\psi \in \Delta$ for all Δ for which s_Δ is G_I-reachable from s_Γ. We have to prove that M-INT$_G(\psi) \in \Gamma$.
First a general remark. Because each s_Δ corresponds to a *finite* set of formulas Δ, each Δ can be represented as the finite conjunction of its formulas, denoted as φ_Δ. Note that it is crucial that we restricted ourselves to the finite closure Φ.
Now define Z as:

$$\{\Lambda \in \text{CON}_\Phi \mid \psi \in \Delta \text{ for all } \Delta \text{ for which } s_\Delta \text{ is } G_I\text{-reachable from} s_\Gamma\}$$

So in particular, $\Gamma \in Z$. Intuitively, Z should become the set of worlds in which M-INT$_G(\psi)$ holds.
Now let:

$$\varphi_Z = \bigvee_{\Lambda \in Z} \varphi_\Lambda$$

This formula is the disjunction of the descriptions of all states corresponding to Z. From the finiteness of Z, it follows that φ_Z is a formula of the language. Similarly, define:

$$\varphi_{\overline{W}} = \bigvee_{\Theta \in \overline{W}} \varphi_\Theta, \text{ where } \overline{Z} = \{\Theta \in \text{CON}_\Phi \mid \Theta \notin Z\}$$

Thus, $\varphi_{\overline{W}}$ can be viewed as the description of all worlds outside Z.

Our aim is to prove the following *Claim*:

$$\text{TEAMLOG}^{\text{mint}} \vdash \varphi_Z \rightarrow \text{E-INT}_G(\varphi_Z)$$

First, let's show how this claim helps to prove the desired conclusion $\text{M-INT}_G(\psi) \in \Gamma$. Because $\psi \in \Lambda$ for all $\Lambda \in Z$ and ψ occurs in all conjunctions φ_Λ for all $\Lambda \in Z$, we have $\text{TEAMLOG}^{\text{mint}} \vdash \varphi_Z \rightarrow \psi$. Starting from this and the claim above, we may distribute E-INT_G over the implication by a number of uses of $\mathbf{R2}_I$, $\mathbf{A2}_I$, $\mathbf{M1}$ and some propositional reasoning to derive:

$$\text{TEAMLOG}^{\text{mint}} \vdash \varphi_Z \rightarrow \text{E-INT}_G(\psi \wedge \varphi_Z)$$

Rule $\mathbf{RM1}$ immediately gives:

$$\text{TEAMLOG}^{\text{mint}} \vdash \varphi_Z \rightarrow \text{M-INT}_G(\psi)$$

Now because φ_Γ is one of the disjuncts of φ_Z, we have:

$$\text{TEAMLOG}^{\text{mint}} \vdash \varphi_\Gamma \rightarrow \text{M-INT}_G(\psi)$$

Finally, using the Consequence Lemma and some more propositional reasoning, we conclude, as desired:

$$\text{M-INT}_G(\psi) \in \Gamma$$

Thus, it remains to prove the claim $\text{TEAMLOG}^{\text{mint}} \vdash \varphi_Z \rightarrow \text{E-INT}_G(\varphi_Z)$. We do this in the following five steps:
1. We first show that for all $i \in G$ and for all $\Lambda \in Z$ and $\Theta \in \overline{W}$:

$$\text{TEAMLOG}^{\text{mint}} \vdash \varphi_\Lambda \rightarrow \text{INT}(i, \neg\varphi_\Theta)$$

So assume that $\Lambda \in Z$ and $\Theta \in \overline{W}$.

By definition of Z and $\overline{W}$, we have $\psi \in \Delta$ for all Δ for which s_Δ is G_I-reachable from s_Λ, but there is a Δ' such that $s_{\Delta'}$ is G_I-reach-able from s_Θ and $\psi \notin \Delta'$. Therefore, $(s_\Lambda, s_\Theta) \notin I_i$ for any $i \in G$. Choose an $i \in G$. By definition of I_i, there is a formula χ_i such that $\text{INT}(i, \chi_i) \in \Lambda$ while $\chi_i \notin \Theta$. As Θ is maximally $\text{TEAMLOG}^{\text{mint}}$-consistent in Φ, we have:

$$\text{TEAMLOG}^{\text{mint}} \vdash \varphi_\Theta \rightarrow \neg\chi_i$$

and thus by contraposition:

$$\text{TEAMLOG}^{\text{mint}} \vdash \chi_i \rightarrow \neg\varphi_\Theta$$

Using $\mathbf{R2}_I$ and $\mathbf{A2}_I$, we derive:

$$\text{TEAMLOG}^{\text{mint}} \vdash \text{INT}(i, \chi_i) \rightarrow \text{INT}(i, \neg\varphi_\Theta)$$

and as $\text{INT}(i, \chi_i) \in \Lambda$, we have:

$$\text{TEAMLOG}^{\text{mint}} \vdash \varphi_\Lambda \rightarrow \text{INT}(i, \neg\varphi_\Theta)$$

2. $\text{TEAMLOG}^{mint} \vdash \varphi_\Lambda \rightarrow \text{INT}(i, \bigwedge_{\Theta \in \overline{W}} \neg\varphi_\Theta)$.
   In fact, this follows from 1 by propositional logic and the well-known derived rule of standard modal logic that intention distributes over conjunctions.

3. Here we show that $\text{TEAMLOG}^{mint} \vdash \bigvee_{\Delta \in \text{CON}_\Phi} \varphi_\Delta$.
   Proof Suppose on the contrary that the formula $\neg\bigvee_{\Delta \in \text{CON}_\Phi} \varphi_\Delta$, which is equivalent by De Morgan's laws to $\bigwedge_{\Delta \in \text{CON}_\Phi} \neg\varphi_\Delta$, is TEAMLOG^{mint}-consistent.
   Then we can find for every $\Delta \in \text{CON}_\Phi$ a conjunct ψ_Δ of φ_Δ such that $\overline{\Delta} := \{\neg\psi_\Delta \mid \Delta \in \text{CON}_\Phi\}$ is TEAMLOG^{mint}-consistent.
   Thus, by Lemma 3.2, there is a set of formulas $\Theta \supseteq \overline{\Delta}$ which is maximally TEAMLOG^{mint}-consistent in Φ. Now we come to the desired contradiction by diagonalization: Θ contains both ψ_Θ (which was defined as a conjunct of φ_Θ) and, because $\Theta \supseteq \overline{\Delta}$, also $\neg\psi_\Theta$.

4. $\text{TEAMLOG}^{mint} \vdash \varphi_Z \leftrightarrow (\bigwedge_{\Theta \in \overline{W}} \neg\varphi_\Theta)$. This follows almost immediately from 3.

5. Here we show the final claim that:

$$\text{TEAMLOG}^{mint} \vdash \varphi_Z \rightarrow \text{E-INT}_G(\varphi_Z)$$

Proof: By 2 and 4 we have for all $i \in G$ that:

$$\text{TEAMLOG}^{mint} \vdash \varphi_\Gamma \rightarrow \text{INT}(i, \varphi_Z)$$

and so by **M1** and some propositional reasoning:

$$\text{TEAMLOG}^{mint} \vdash \varphi_\Gamma \rightarrow \text{E-INT}_G(\varphi_Z)$$

Finally, because $\Gamma \in Z$, our claim holds.

Lemma 3.5 (Finite Truth Lemma) *If $\Gamma \in \text{CON}_\Phi$, then for all $\psi \in \Phi$ it holds that $M_\varphi, s_\Gamma \models \psi$ iff $\psi \in \Gamma$.*

Proof Immediately from the Finite Valuation Lemma, by induction on the structure of ψ. Details are left to the reader.

Theorem 3.1 (Completeness of TEAMLOG^{mint}) *If $\text{TEAMLOG}^{mint} \nvdash \varphi$, then there is a serial model M and a $w \in M$ such that $M, w \nvDash \varphi$.*

Proof Assume that $\text{TEAMLOG}^{mint} \nvdash \varphi$. Take M_φ as defined in definition 3.8. Note that there is a formula χ logically equivalent to $\neg\varphi$ that is an element of Φ; if φ does not start with a negation, χ is the formula $\neg\varphi$ itself. Now, using Lemma 3.1, there is a maximally consistent $\Gamma \subseteq \Phi$ such that $\chi \in \Gamma$. By the Finite Truth Lemma, this implies that $M_\varphi, s_\Gamma \models \chi$, thus $M_\varphi, s_\Gamma \nvDash \varphi$. The model is serial, because for all $s_\Delta \in S_\varphi$ we have by the Finite Valuation Lemma that $M, s_\Delta \models \neg\text{INT}(i, \bot)$ for all $i \leq m$; so all worlds have I_i successors for all agents.

The full presentation of the proof is meant to suggest to the reader that the method may be adapted to prove completeness as well for the combined systems for individual and common beliefs and mutual intentions such as $\text{KD}_n^{\text{C-INT}_G}$ and TEAMLOG as a whole. Also dynamic logic may be added. In Chapter 9, alternative proof methods for decidability by semantic tableaux are given for these systems and those proofs have completeness as their by-product.

As a further example, the theory $\mathrm{KD}_n^{\text{M-INT}'_G}$ of Section 3.7 can be seen to be sound and complete with respect to Kripke models where all n accessibility relations for both I and B are serial, while those for B are additionally transitive and Euclidean, by adaptation of the proof above.

As to complexity of the decidability for the logics introduced in this chapter, TEAMLOG^{ind} turns out to be PSPACE-complete, just like its component individual modal logics. The encompassing logic for teamwork TEAMLOG, on the other hand, is EXPTIME-complete, due to the recursive character of the collective notions. These issues will be further discussed in Chapter 9.

3.9   Related Approaches to Intentions in a Group

The most influential theory of teamwork is the one of Wooldridge and Jennings (1999). The actual formal frameworks of their papers is quite different from ours. Wooldridge and Jennings (1999) define joint commitment towards φ in a more dynamic way than we define collective intentions: initially, the agents do not believe φ is satisfied ($\neg\text{BEL}(i, \varphi)$), and subsequently have φ as a goal until the termination condition is satisfied, including (as conventions) conditions on the agents to turn their eventual beliefs that termination is warranted into common beliefs. Subsequently, they define having a joint intention to do α as 'having a joint commitment that α will happen next, and then α happens next'. In contrast, agreeing with Castelfranchi (1995), we view collective commitments as stronger than collective intentions and base the collective commitment on a specific social plan meant to realize the collective intention. Our ideas on collective commitments are presented in Chapter 4 as well as in Chapter 6, which discusses the dynamic aspects.

The emphasis on establishing appropriate collective attitudes for teamwork is shared with the SharedPlans approach of Grosz and Kraus (1996, 1999). Nevertheless, the intentional component in their definition of collective plans is weaker than our collective intention: Grosz and Kraus' agents involved in a collective plan have individual intentions towards the overall goal and a common belief about these intentions; intentions with respect to the other agents play a part only at the level of individual subactions of the collective plan.

We stress, however, that team members' intentions about their colleagues' motivation to achieve the overall goal play an important role in keeping the team on track even if their plan has to be changed radically due to a changing environment (see Chapters 5 and 6).

Balzer and Tuomela (1997) take a technical approach using fixed points, inspired by the work on common knowledge in epistemic logic (Fagin et al., 1995; Meyer and van der Hoek, 1995). They define we-attitudes such as collective goals and intentions using fixed-point definitions. Our definitions use fixed-point constructions as well, but interpret collective intentions a bit differently. In Balzer's and Tuomela's view, abilities and opportunities play a part during the construction of a collective intention (the stage of team formation). In our approach, on the other hand, abilities are mainly important at the two surrounding stages, namely during potential recognition (before the stage of team formation) and during plan formation, where a collective commitment is established on the basis of a collective intention and a social plan (see Chapters 5, 6 and 8).

Rao *et al.* (1992) consider some related issues with an emphasis on the ontology and semantics of social agents carrying out social plans. They use a much weaker definition of joint intention than ours: it is only one-level, being defined as 'everyone has the individual intention, and there is a common belief about this'. Thus, their definition does not preclude cases of coercion and competition.

Haddadi (1995) gives an internal or prescriptive approach that characterizes the stages of cooperative problem solving in a manner similar to Wooldridge and Jennings (1999), but is based on the branching-time semantics of Rao and Georgeff (1991) instead of the linear-time semantics of Levesque *et al.* (1990). She introduces the notions of pre-commitments and commitments between pairs of agents and presents an extensive and well-founded discussion of their properties, including important aspects like communication. However, in contrast to our approach, she does not go beyond the level of pairwise commitments and is not explicit about their contribution to collective behavior in a bigger team.

An early alternative account of group intentions was given by Singh (1990). He criticizes two theoretical proposals about group intentions that were originally proposed to model human discourse, namely the SharedPlans approach of *Plans for discourse* (Grosz and Sidner, 1990) and the work *On acting together* (Levesque *et al.*, 1990), arguing that they are not suited for modeling more general types of cooperation in distributed artificial intelligence. Singh notes that in these two theories, common belief is an integral part of a group's intention, which may be problematic because creating common beliefs is costly in terms of communication and impossible if the communication medium is untrustworthy; we have discussed these problems in Section 2.7 and Section 3.7. Moreover, Singh objects that Levesque *et al.* (1990) posit the obligation to communicate about dropping intentions as part and parcel of the concept of group intentions, while this is a convention that holds in some contexts but not in others. Finally, Singh deplores the assumption of a homogeneous group and the absence of social structure in the two accounts. Singh's own solution is to model group intention based on branching time temporal logic, where the main ingredients are *strategies* performed by a group and where the social structure of the group is taken into account (Singh, 1990, 1998). We agree that accounting for a group's social structure is crucial in a theory of teamwork and in fact we incorporate it in our investigation of collective commitments in the next chapter.

Collective intentions and collective commitments do appear on center stage in the work of Margaret Gilbert, who has developed a philosophical theory of the plural subject since her seminal book *On Social Facts* (Gilbert, 1989). Collective informational and motivational attitudes play a very important role also in her later work; for example, a nice survey about her view on their role in teamwork is presented in Gilbert (2005). Even though her research is strictly philosophical, it offers fruitful inspiration for future work on teamwork in multi-agent systems. For example, it would be interesting to formalize her ideas on whether there is any such thing as collective responsibility (Gilbert, 2009). Especially in the context of collective commitments, treated in the next chapter, one may build on her investigations into the ways in which agreements lead to obligations.

3.9.1 What Next?

On the basis of individual characteristics of particular agents, their mutual dependencies and other possibly complex criteria, one can classify and investigate different types of teams, various types of cooperation, communication, negotiation, etc. An interesting extension to other than strictly cooperative groups is an interesting subject of future research.

Although we view collective intention as a central concept during the whole process of teamwork, in this present chapter we focus on its static aspects during planning. In general, the presented definitions of collective intentions in terms of more basic attitudes may be combined with either dynamic or temporal logic, depending on the application. The proper treatment of collective intentions, as well as commitments, in a dynamically changing environment entails the maintenance of all individual, social and collective motivational attitudes involved throughout the whole process. In Chapter 5, a generic reconfiguration algorithm for BDI systems is presented. In Chapter 6, we investigate the persistence and evolution of motivational attitudes during teamwork. In addition, in Chapter 8 we characterize the role of dialogue in teamwork.

4

A Tuning Machine for Collective Commitments

Stop trying to control.
Let go of fixed plans and concepts,
and the world will govern itself.

Tao Te Ching (Lao-Tzu, Verse 57)

4.1 Collective Commitment

4.1.1 Gradations of Teamwork

The commonsense meaning of teamwork covers different gradations of being a team. Take, as a first example, teamwork in a group of researchers who jointly plan their research and divide roles, who reciprocally keep a check on how the others are doing and help their colleagues when needed in furtherance of their goal to prove a theorem. All aspects of teamwork are openly discussed in the team, and members keep one another informed about changes in the plan. Therefore, this is a paradigmatic example of teamwork.

Contrast this kind of non-hierarchical teamwork with a second example, of a group of spies who all work on the same goal, say to locate agent X. In their case a plan is designed by one mastermind, who divides the roles and divulges to each participant *only* the necessary information. Thus, members may not even know the main goal, nor who else is included in the group. Even though the connection between group members is rather loose, we would still like to speak about Cooperative Problem Solving (henceforth CPS), albeit a non-typical case, and not about proper teamwork.

In the examples, individual and group awareness about such ingredients of CPS like the main goal and the plan to achieve it, range from very high (as in the first example above) to very low (as in the second example). Therefore, we claim that these two cases cannot be reasonably covered by *one* generic logical model of teamwork. Thus far in the MAS literature, authors restricted themselves to a typical idealized understanding of teamwork, usually abstracting from organizational structures and communication possibilities

Teamwork in Multi-Agent Systems: A Formal Approach Barbara Dunin-Kęplicz and Rineke Verbrugge
© 2010 John Wiley & Sons, Ltd

(Dunin-Kęplicz and Verbrugge, 1996; Grosz and Kraus, 1999; Rao *et al.*, 1992; Wooldridge and Jennings, 1999). In contrast, in the sequel we will provide a full characterization of group attitudes, covering the range from proper teams to more loosely connected groups involved in CPS. We will also highlight the importance of choosing an appropriate gradation of teamwork needed for a specific goal in given circumstances. Therefore, a mechanism will be provided to create an adequate type of commitment. The proposed model of collective commitments is minimal, in order to support the system developer's quest for efficiency at design time.

4.1.2 Collective Commitment Triggers Team Action

Suppose we have a team with a collective intention to achieve a goal φ. Does this suffice for the team to start its cooperative action towards the goal? Clearly not: a bridge from a still rather abstract collective intention to precise team action is needed.

What is obviously missing is a detailed plan including individual actions to realize the goal. This would enable the agents to make bilateral 'promises', called social commitments, to perform their parts. Why bilateral? Because by their simplicity, they are easy to implement and revise. The collective motivational attitude that covers the outcome of this planning and committing is a team's collective commitment or a weaker attitude that plays a similar cohesive role. Ultimately, collective commitment in a group of agents is aimed to trigger team action, that is, a coordinated execution of agents' individual actions according to the adopted social plan. A formal model of a group's motivational stance towards teamwork is the focus of this chapter.

Again, agents' *awareness* about the overall situation is vital. As a reminder, the notion of awareness applied in modeling agency may be viewed as a reduction of a general sense of 'consciousness' to an agent's beliefs about itself, about other agents and finally about the state of an environment (as discussed in Section 2.7), naturally expressed by different degrees of beliefs. These range from the rather strong common beliefs through weaker forms, like possibly iterated general belief, to even weaker individual beliefs, depending on the circumstances.

4.1.3 A Tuning Mechanism

When asking what it means for a group of agents to be *collectively committed* to achieve a common goal, both the circumstances in which the group is acting and the structure of the organization it is a part of, have to be taken into account. This implies the importance of differentiating the scope and strength of the group commitment. The resulting characteristics may differ significantly and even become logically incomparable.

The idea of *dials* to tune the nature of the commitment to the particular purpose seems to be both technically interesting and intuitively appealing. We intend to provide a sort of *tuning mechanism* which enables the system developer to *calibrate* a type of collective commitment fitting the circumstances, analogously to adjusting dials on an audio system. The appropriate dials, characterized in the sequel, belong to a device representing a general schema of collective commitment.

In order to illustrate the expressive power of such a *tuning machine*, several types of commitments corresponding to various teamwork schemes occurring in practice will be discussed. Apparently, the entire spectrum of possibilities is much wider, due to the number of possibly independent choices to be made. The resulting types of collective or group commitments, described in multimodal logic, may then be naturally implemented in a specific multi-agent system. In this way the tuning mechanism may be viewed as a bridge between theory and practice.

In this chapter we concentrate on a static theory of teamwork defining complex social and collective motivational attitudes in terms of simpler individual ones. The next three chapters, in contrast, focus on the dynamics of individual intention and social commitment in the context of cognitive and social processes involved (see also Castelfranchi (1999), Dignum and Conte (1997) and Dignum *et al.* (2001b)).

The rest of the chapter is structured in the following way. In Sections 4.2 and 4.3, a short presentation is given of the logical framework extending that of the previous two chapters in order to construct the building blocks of collective commitments. The central Sections 4.4 and 4.5 explore different dimensions along which collective commitments may be tuned to fit both the organization and the environment. A general scheme is presented in a multi-modal language, as well as five different notions of collective commitment fitting to concrete organizational structures. Section 4.6 explores how several interesting organizational topologies, such as stars, rings and trees, can be explicitly represented in definitions of collective commitment. Finally, Section 4.7 focuses on discussion and provides a bridge to the subsequent chapters about the dynamic part of the story of teamwork. The reader may skip Sections 4.2 and 4.3 at first reading, and instead start reading from Section 4.4, only jumping back when needing more background about the building blocks of collective commitment.

4.2   The Language and Kripke Semantics

We propose the use of multi-modal logics to formalize agents' motivational attitudes as well as actions they perform and their effects.

4.2.1   Language

Individual actions and formulas are defined inductively, both with respect to a fixed finite set of agents. The basis of the induction is given in the following definition.

Definition 4.1 (Basic elements of the language)   The language is based on a denumerable set $\mathcal{P}$ of *propositional symbols* and a finite set $\mathcal{A}$ of *agents*, as in definition 2.1, extended with:

- a finite set $\mathcal{A}t$ of *atomic actions*, denoted by a or b.

In TEAMLOG, most modalities relating agents' motivational attitudes appear in two forms: with respect to *propositions*, or with respect to *actions*, the choice depending on the context. These actions are interpreted in a generic way – we abstract from any particular form of actions: they may be complex or primitive, viewed traditionally with certain effects or with default effects (Dunin-Kęplicz and Radzikowska, 1995a, b, c), etc.

A proposition reflects a particular state of affairs. The transition from a proposition that an agent aims for to an action realizing this, is achieved by means-end analysis. The set of formulas is defined by simultaneous induction, together with the sets of individual actions and social plan expressions (see Definitions 4.3 and 4.4). It is extended with other needed modalities. These have all been explained where they first appeared in this book. See Section 2.4 about epistemic modalities and Sections 3.3, 4.3.2 and 3.5 about individual, social and collective motivational modalities with the following additional inductive clauses.

Definition 4.2 (Formulas) We inductively define a set of formulas $\mathcal{L}$, an extension of the language given in Definitions 2.2 and 3.1.

F5 If φ is a formula, α is an individual action, $i, j \in \mathcal{A}$, $G \subseteq \mathcal{A}$, and P a social plan expression, then the following are formulas:
motivational modalities
GOAL(i, α), INT(i, α); COMM(i, j, φ), COMM(i, j, α);
E-INT$_G(\varphi)$, E-INT$_G(\alpha)$, M-INT$_G(\varphi)$, M-INT$_G(\alpha)$, C-INT$_G(\varphi)$, C-INT$_G(\alpha)$;
R-COMM$_{G,P}(\varphi)$, R-COMM$_{G,P}(\alpha)$, S-COMM$_{G,P}(\varphi)$, S-COMM$_{G,P}(\alpha)$;
W-COMM$_{G,P}(\varphi)$, W-COMM$_{G,P}(\alpha)$, T-COMM$_{G,P}(\varphi)$, T-COMM$_{G,P}(\alpha)$;
D-COMM$_{G,P}(\varphi)$, D-COMM$_{G,P}(\alpha)$.

Table 4.1 gives a number of new formulas additional to those of Table 2.1 of Chapter 2, and Table 3.1 and 3.2 of Chapter 3. The formulas in this table concern motivational attitudes appearing in this chapter with their intended meanings. The symbol φ denotes a proposition and all notions also exist with respect to an action α. Thus, the action notions E-INT$_G(\alpha)$, M-INT$_G(\alpha)$ and C-INT$_G(\alpha)$ are governed by axioms and semantics analogous to those given in Chapter 3 for the versions with respect to propositions and we do not explicitly state them here.

Table 4.1 New formulas and their intended meanings.

INT(i, α)	Agent i has the intention to do α
E-INT$_G(\alpha)$	Every agent in group G has the individual intention to do α
M-INT$_G(\alpha)$	Group G has the mutual intention to do α
C-INT$_G(\alpha)$	Group G has the collective intention to do α
COMM(i, j, φ)	Agent i commits to agent j to achieve φ
COMM(i, j, α)	Agent i commits to agent j to do α
R-COMM$_{G,P}(\varphi)$	Group G has robust collective commitment to achieve φ by plan P
R-COMM$_{G,P}(\alpha)$	Group G has robust collective commitment to do α by plan P
S-COMM$_{G,P}(\varphi)$	Group G has strong collective commitment to achieve φ by plan P
S-COMM$_{G,P}(\alpha)$	Group G has a strong collective commitment to do α by plan P
W-COMM$_{G,P}(\varphi)$	Group G has weak collective commitment to achieve φ by plan P
W-COMM$_{G,P}(\alpha)$	Group G has a weak collective commitment to do α by plan P
T-COMM$_{G,P}(\varphi)$	Group G has team commitment to achieve φ by plan P
T-COMM$_{G,P}(\alpha)$	Group G has team commitment to do α by plan P
D-COMM$_{G,P}(\varphi)$	Group G has distributed commitment to achieve φ by plan P
D-COMM$_{G,P}(\alpha)$	Group G has distributed commitment to do α by plan P

We will subsequently define the set of individual actions $\mathcal{A}c$, and the set of social plan expressions $\mathcal{S}p$, combining individual actions into group ones. Below, we give a particular choice of operators to define individual actions and social plan expressions. However, in the sequel we hardly come into detail as to how particular individual actions and social plans are built up. Thus, another definition (for example without the iteration operation or without non-deterministic choice) may be used if more appropriate in a particular application.

Definition 4.3 (Individual actions)   The set $\mathcal{A}c$ of *individual actions* is defined inductively as follows:

AC1 each atomic action $a \in \mathcal{A}t$ is an individual action;

AC2 if $\varphi \in \mathcal{L}$, then $\texttt{confirm}\varphi$ is an individual action;   (Confirmation)

AC3 if $\alpha_1, \alpha_2 \in \mathcal{A}c$, then $\alpha_1; \alpha_2$ is an individual action;   (Sequential Composition)

AC4 if $\alpha_1, \alpha_2 \in \mathcal{A}c$, then $\alpha_1 \cup \alpha_2$ is an individual action;   (Non-Deterministic Choice)

AC5 if $\alpha \in \mathcal{A}c$, then α^* is an individual action;   (Iteration)

AC6 if $\varphi \in \mathcal{L}$, then $\texttt{stit}(\varphi)$ is an individual action.

Here, in addition to the standard dynamic operators of **AC1** to **AC5**, the operator $\texttt{stit}$ of **AC6** stands for 'sees to it that' or 'brings it about that' and has been extensively treated in Segerberg (1989).

Definition 4.4 (Social plan expressions)   The set $\mathcal{S}p$ of social plan expressions is defined inductively as follows:

SP1 If $\alpha \in \mathcal{A}c$ and $i \in \mathcal{A}$, then $\langle \alpha, i \rangle$ is a well-formed social plan expression;

SP2 If α and β are social plan expressions, then $\langle \alpha;\beta \rangle$ (sequential composition) and $\langle \alpha \parallel \beta \rangle$ (parallelism) are social plan expressions.

A concrete example of a social plan expression will be given in Section 4.3.1.

4.2.2   Kripke Models

Each Kripke model for the language defined in the previous subsection consists of a set of worlds, a set of accessibility relations between worlds and a valuation of the propositional atoms, as given in Definition 3.2.

Definition 4.5 (Kripke model)   A Kripke model is a tuple
$\mathcal{M} = (W, \{B_i : i \in \mathcal{A}\}, \{G_i : i \in \mathcal{A}\}, \{I_i : i \in \mathcal{A}\}, Val)$, such that:

1. W is a set of possible worlds, or states.
2. For all $i \in \mathcal{A}$, it holds that $B_i, G_i, I_i \subseteq W \times W$. They stand for the accessibility relations for each agent with respect to beliefs, goals and intentions, respectively.

3. *Val* : $\mathcal{P} \times W \rightarrow \{0, 1\}$ is the function that assigns the truth values to propositional formulas in states.

The truth conditions for the propositional part of the language are all standard. Those for the modal operators are treated in Chapter 2 (the informational attitudes) and Chapter 3 (goals and intentions with respect to propositions). Here follows the adapted definition for individual motivational attitudes with respect to actions.

Definition 4.6 (Truth definition)

- $\mathcal{M}, s \models \text{INT}(i, \alpha)$ iff $\mathcal{M}, t \models done(i, \alpha)$ for all t such that $s I_i t$.

Here, $done(i, \alpha)$ means that agent i has just performed action α. We do not want to come into details about dynamic and temporal aspects here but see Chapter 6 for semantics of $done(i, \alpha)$ in a dynamic logic framework.

4.3   Building Collective Commitments

What is the minimal set of ingredients that are essential for building a viable notion of group commitment? In our investigation, we have isolated and separately characterized three important ingredients of group commitments. In addition, a group may stay aware of these aspects in different ways:

1. *A group* – usually, a strictly cooperative team has to be formed. In TEAMLOG, a team is established on the basis of a *collective intention*.
2. *A plan* – a *social plan* that details how to realize the group's goal needs to be created.
3. *A distribution of responsibilities* – a set of pairwise social commitments towards the actions from the social plan reflects the agents' responsibilities during team action.

As collective intentions have been extensively treated in the previous chapter, we will focus on the remaining ingredients, starting from social plans.

4.3.1   Social Plans

Collective commitments are plan-based: they are defined with respect to a given *social plan*. Individual actions (from $\mathcal{A}c$, see Section 4.2) may be combined into group actions by *social plan expressions*, as in Definition 4.4 of Section 4.2. The social plan should be effective, as reflected in the predicate *constitute*(φ, P), to be explained in Section 6.4.1. This predicate states that successful realization of the plan P leads to the achievement of goal φ.

Let us give a simple example of a social plan, based on the first example in Section 4.1. Consider a team consisting of three agents t (the theorem prover), l (the lemma prover) and c (the proof checker) who have a collective intention to prove a new mathematical theorem. In joint deliberation, they have divided their roles according to their abilities and preferences. Suppose during planning they formulate two lemmas, which still need to be proved, and the following complex individual actions: *provelemma1*, *provelemma2* (to prove lemma 1, respectively 2), *checklemma1*, *checklemma2* (to check a proof of lemma 1, respectively 2), *provetheorem* (prove the theorem from the conjunction of lemmas 1 and 2) and *checktheorem* (to check the proof of the theorem from the lemmas). One possible social plan they can come up with is the following. First, the lemma prover, who proves lemmas 1 and 2 in succession, and the theorem prover, who proves the theorem from the two lemmas, work in parallel, and subsequently the proof checker checks their proofs in a fixed order, formally:

$$P = \langle\langle\langle\langle provelemma1, l\rangle; \langle provelemma2, l\rangle\rangle \parallel \langle provetheorem, t\rangle\rangle;$$

$$\langle\langle\langle checklemma1, c\rangle; \langle checklemma2, c\rangle\rangle; \langle checktheorem, c\rangle\rangle\rangle$$

Consider again this group of agents with the same collective intention to prove a new mathematical theorem. In the course of planning they formulate two lemmas, but this time either one of the lemmas suffices to prove the theorem as follows:

$$P = \langle\langle\langle\langle\langle provelemma1, l\rangle; \langle checklemma1, c\rangle\rangle \cup \langle\langle provelemma2, l\rangle ;$$

$$\langle checklemma2, c\rangle\rangle\rangle; \langle provetheorem, t\rangle\rangle; \langle checktheorem, c\rangle\rangle$$

We will use this context as a running example in Section 6.5.

4.3.2 Social Commitments

In our model of teamwork, pairwise or social commitments are first-class citizens. Most of the time, cooperation between two agents involves a certain asymmetric role division: the first agent (called j) wants some state of affairs to be achieved (an action to be performed) while a second agent (called i) decides that it can perform the action needed. When j is willing to have i as a helper and to oversee the achievement of the goal (the performance of the action), they recognize their potential for cooperation. This recognition is reflected in a promise from i to j. A *social commitment* is the bilateral motivational attitude that corresponds to such a promise.

Thus, a social commitment understood this way is stronger than an individual intention. If i commits to j to do something, then in the first place i has the *intention* to do that. Moreover, j should be *interested* in i fulfilling its intention. These two conditions (inspired by Castelfranchi, 1995), need to be enhanced by the condition expressing the agents' awareness about the situation, that is about their individual attitudes.[1] Such awareness is generally achieved by communication. In our earlier papers, awareness was expressed in terms of common belief (Dunin-Kęplicz and Verbrugge, 2004). In the sequel, social commitments are characterized using an *awareness$_G$*-dial. Two characterizations are given, with respect to actions α and propositions φ, respectively:

$$\text{COMM}(i, j, \alpha) \leftrightarrow \text{INT}(i, \alpha) \wedge \text{GOAL}(j, done(i, \alpha)) \wedge$$

$$awareness_{\{i,j\}}(\text{INT}(i, \alpha) \wedge \text{GOAL}(j, done(i, \alpha)))$$

$$\text{COMM}(i, j, \varphi) \leftrightarrow \text{INT}(i, \varphi) \wedge \text{GOAL}(j, done(i, \texttt{stit}(\varphi)) \wedge$$

$$awareness_{\{i,j\}}(\text{INT}(i, \varphi) \wedge \text{GOAL}(j, done(i, \texttt{stit}(\varphi)))$$

---

[1] See also Searle (1969) for an early discussion about the properties of promises.

Here, $done(i, \texttt{stit}(\varphi))$ stands for "agent i has seen to it that φ". Indeed, $\texttt{stit}$ can be seen as a shortcut, turning the achievement of a state of the world reflected by proposition φ, into a possibly complex action. To formalize it, one can move to a second-order language to quantify over scenarios. Alternatively, Horty and Belnap (1995) propose a branching time framework. More recently, quantification over possible strategies has found a natural expression in Alternating Time Temporal Logic (ATL), which can be embedded in a logic for strategic $\texttt{stit}$ (Broersen et al., 2006). We do not want to tie ourselves to a specific choice of formalization and do not analyze $\texttt{stit}$ any further.

If the *awareness$_G$*-dial is placed at C-BEL$_G$ for $G = \{i, j\}$, then social commitment obeys positive introspection, namely:

$$\text{COMM}(i, j, \varphi) \rightarrow \text{C-BEL}_G(\text{COMM}(i, j, \varphi))$$

This follows from the awareness condition included in the definition. Note that it is not possible to derive negative introspection, because agents are in general not aware of the absence of common beliefs (that is $\neg\text{C-BEL}_G(\varphi) \rightarrow \text{BEL}(i, \neg\text{C-BEL}_G(\varphi))$ is not provable for $i \in G$).

The above definitions present the bare ingredients of social commitments and not the process leading to their establishment. It may happen that the language of informational and motivational attitudes is not sufficiently fine-grained to express various subtle aspects involved. In fact, both *causality* and *obligation* come to the fore when creating social commitments. Usually agent i takes on a social commitment $\text{COMM}(i, j, \alpha)$ *because* the other agent is interested in it, even though such causality is not explicitly reflected in the definition (see Castelfranchi, 1999 for a recent discussion of causality and commitment). Then after adoption, social commitments lead to an obligation for agent i to fulfill its promise of performing the action or achieving the goal. In the definition, only the final formal outcome is shown, as befits the static part of our theory of teamwork. The more dynamic, process-oriented part TEAMLOG^{dyn} of the story will be told in Chapters 5 and 6.

4.3.3 Deontic Aspects of Social Commitments

Castelfranchi (1995) states that 'if I commit to you to do something, then I *ought* to do it'. Additionally, we find that the strength of the obligation depends on the situation and the agents, for example, are the agents involved responsible ones? Below we give an axiom that characterizes responsible agents by relating social commitments and obligations. It reflects our view that the obligation is related to the current state of the agent's commitment: only as long as an agent's commitment is still valid, it ought to fulfill it. Formally:

$$\text{COMM}(i, j, \varphi) \rightarrow \mathbf{A}\,(\text{OUGHT}(i, \varphi)\,\mathbf{U}\,\neg\text{COMM}(i, j, \varphi))$$

Here, $\text{OUGHT}(a, \varphi)$ is a modal operator with intended reading 'i is obliged to achieve φ'. The axiom is formulated in the temporal language and means informally: 'If i commits to j to achieve φ, then i is obliged to achieve φ until its social commitment has been dropped appropriately'. The axiom above is meant only as a possible extension of TEAMLOG.

There are many axiom systems and corresponding semantics in the literature on deontic logic (cf. Aaqvist, 1984). There are also systems in which agents have obligations not merely towards propositions, but also with respect to actions (cf. d'Altan *et al.*, 1993).

It is sufficient, though, to assume a standard propositional KD-type modal logic for the obligations of each agent. In the corresponding Kripke semantics, there are the usual accessibility relations R_a that lead from worlds w to worlds that are 'optimal' for agent a in w. These accessibility relations are serial, corresponding to the consistency of obligations.

4.3.4 Commitment Strategies

Let us peak into agents' differing propensities to maintain or drop their social commitments. The key point is whether and in which circumstances an agent is allowed to drop a social commitment. If such a situation arises, the next question is how to deal with it responsibly. The definitions are inspired by those of Rao and Georgeff (1991) for intention strategies. The need for agents' responsible behavior led us to include additionally the social aspects of communication and coordination. We assume that the commitment strategies are an immanent property of the individual agent and that they do not depend on the goal to which the agent is committed, nor on the other agent to whom it is committed. We also assume that each agent knows which commitment strategies are adopted by all agents in the group. This 'meta-knowledge' ensures proper replanning and coordination (Dunin-Kęplicz and Verbrugge, 1996). See the Appendix for the formal definitions and Dunin-Kęplicz and Verbrugge (1999) and Rao and Georgeff (1991) for more discussion.

The strongest commitment strategy is followed by the *blindly committed* agent, who maintains its commitments until it actually believes that they have been achieved.

Single-minded agents may drop a social commitment when they do not believe anymore that it is realizable. However, as soon as the agent abandons a commitment, it needs to communicate and coordinate with the agent to whom it is committed.

For *open-minded* agents, the situation is similar as for single-minded ones, except that they can also drop social commitments if they do not aim for the respective goal anymore. As in the case of single-minded agents, communication and coordination should be involved.

There still remains the important problem of the consequences of an open-minded agent dropping a social commitment. We assume here that it is allowed to do this after communicating and coordinating with its partner. This solution, however, is not always subtle enough. We agree with Castelfranchi (1995) and Tuomela (1995) that in some cases dropping a social commitment should be more difficult and should cause real consequences for an agent, potentially expressed in extra axioms.

When analyzing the possibilities of cooperation, it turns out that blindly-committed agents, who seem most trustworthy at first sight are hard to cooperate with when replanning is needed (Dunin-Kęplicz and Verbrugge, 1996). Thus, the distribution of commitment strategies in a team appears to be important.

Of course it is possible to classify agents along different lines: for example one may characterize them according to eagerness to adopt new social commitments (see Cavedon *et al.*, 1997) etc.

4.4 Tuning Collective Commitments

4.4.1 Why Collective Commitment?

While defining complex motivational attitudes from simpler informational and motivational ones, we view a *social commitment*, not an intention (as in the Rao and Georgeff

(1991) framework), as the trigger for action. In this way we follow the linguistic tradition that distinguishes intentions and commitments.

When investigating collective commitments, we share with Gilbert (2005) her view on its central role in social settings:

> My belief in the importance of joint commitment was the result of extended reflection on the nature of a number of social phenomena referred to in everyday discourse. These phenomena included social conventions and rules, the so-called beliefs of groups, group languages. collective or shared actions, and social groups themselves.

She goes on to provide a number of important aspects of joint commitments (Gilbert, 2005):

1. A joint commitment is not constituted by nor does it entail a set of personal commitments (. . .)
2. Nonetheless, it has implications for the individual participants: *each is committed through it*.
3. Each party is *answerable to all of the parties* for any action that fails to conform to the joint commitment. This is a function of its *jointness*. In the case of failure each can say to the other: you have not simply failed to conform to a commitment of your own. You have failed to conform to *our* commitment, a commitment in which I, as well as you, have an integral stake.
4. People jointly commit to *doing something as a body*, where 'doing something' is construed broadly so as to include, for instance, intending and believing. (. . .)
5. All of the parties must play a role in the creation of a given joint commitment. Given special *'authority-creating' side understandings*, some particular person or body consisting of fewer than all may create a joint commitment for all the parties.
6. *Without* special side understandings, no individual party to a joint commitment can rescind a joint commitment unilaterally.

In our approach to collective commitments, we attempt to ensure desiderata 1–4 in this chapter, while desiderata 5–6 are kept in mind in the dynamic investigations of Chapters 5 and 6. Let us now develop our *plan-based* view of collective commitments, keeping in mind its application in multi-agent settings.

While a collective intention constitutes a team, the collective commitment reflects the concrete manner of achieving the goal by the team. This is provided by planning and hinges on the allocation of actions from a social plan. This allocation is concluded when agents accept pairwise (that is social) commitments to realize their individual actions. Ultimately, a plan-based collective commitment triggers team action. This procedure is generic, so do we strive for a unique generic definition covering many real-life situation? The short answer is 'No'. Instead of creating a possibly too generic notion badly fitting a variety of situations, we design a tuning mechanism to calibrate the scope of collective commitment.

What elements will be the subject of this tuning then? Our experience in modeling group behavior shows that it is agents' awareness that forms the main point of difference over various contexts of common activity. In short, in teamwork awareness concerns the question *who* needs to know *what* in order to cooperate effectively? Actually, there is no point in a generic answer to this question. On the contrary, we look for minimal solutions per context, because the communication and reasoning processes necessary for higher levels of awareness are costly and complex. This is especially important when time is critical. Moreover, awareness of different aspects of teamwork should be calibrated separately in order to achieve the highest possible efficiency and effectiveness. Therefore the separate 'dials' for tuning various aspects of collective commitment constitute a significant part of the tuning mechanism.

4.4.2 General Schema of Collective Commitment

In our generic description we will solely define the basic ingredients constituting collective commitments, leaving room for case-specific extensions. We have recognized the following obligatory ingredients related to different aspects of teamwork:

1. Mutual intention $M\text{-}INT_G(\varphi)$ between a group of agents, allowing them to act as a team (see Section 3.5 for a formal definition and discussion).
   The team exists as long as the mutual intention between team members exists. Thus, no teamwork is considered without a mutual intention among team members.
2. Social plan P for a team on which a collective commitment is based (see Section 4.3.1 for an example).
   The social plan provides a concrete manner to achieve a common goal, the object of mutual intention. Furthermore, plan P should correctly lead to achievement of goal φ, as reflected in $constitute(\varphi, P)$ (see Section 6.4.1).
3. Pairwise social commitments $COMM(i, j, \alpha)$ for actions occurring in the social plan (for a definition of social commitments, see Section 4.3.2).
   Actions from the plan are distributed over team members who accept corresponding social commitments.

Different degrees of awareness about these obligatory ingredients are represented by different dials to be tuned separately. Such a dial may range from the lack of any awareness to common belief. Let $awareness^i_G(\psi)$ stand for 'group G is aware that ψ', where $i = 1, 2, 3$, expressing the group's type of awareness of each of the three above aspects of collective commitment.

4.4.2.1 Detailed versus Global Awareness

The $awareness^3_G$-dial concerns the distribution of actions and social commitments between pairs of team members. In this way a social structure is created and the plan acquires the property of being social. We make a quite refined distinction here, expressing the important difference between *detailed* versus *global* group awareness. This difference is inspired by the *de re* / *de dicto* distinction stemming from the philosophy of language (Quine, 1956). Let us give a short explanation.

The sentence 'Alice is looking for a unicorn' is clearly ambiguous. One reading, called *de re*, is that there is a particular unicorn for which Alice is looking. This reading implies that unicorns exist; then the sentence could be followed by 'but she won't be able to find it'. The other reading, called *de dicto*, is weaker in that it relates Alice to the concept of unicorn and it does not imply that unicorns actually exist. In this reading, the sentence could be followed by 'but she won't be able to find one'. See Montague (1973) for a classic formal treatment of this distinction and many more examples concerning unicorns.

Let us apply the distinction de re / de dicto to the context of informational and motivational attitudes. For example, the sentence 'there is an object of which j believes that it has property A' ($\exists x \text{BEL}(j, A(x))$) is a *de re* belief attribution, which relates agent j to a *res*, namely an individual that the belief is about. On the other hand, 'j believes that there is an object with property A' ($\text{BEL}(j, \exists x A(x))$) is a *de dicto* belief attribution, relating agent j to a *dictum*, namely the proposition $\exists x A(x)$.[2] This distinction is also fruitful for complex epistemic operators such as common belief and the other kinds of awareness used here.

Now we are ready to distinguish the two kinds of group awareness about the social commitments:

1. A *detailed* collective awareness of each social commitment:
   $\bigwedge_{\alpha \in P} \bigvee_{i,j \in G} awareness_G^3(\text{COMM}(i, j, \alpha))$
   In words, for every action α from social plan P, there is a pair of agents i and j such that the group G is aware that agent i is socially committed to agent j to fulfill action α. This corresponds to the interpretation *de re*.
2. A *global* collective awareness of the bare existence of social commitments:
   $awareness_G^3(\bigwedge_{\alpha \in P} \bigvee_{i,j \in G} \text{COMM}(i, j, \alpha))$
   In words, the group G is aware that for every action α from social plan P, there is a pair of agents i and j such that agent i is socially committed to agent j to fulfill action α. This corresponds to the interpretation *de dicto*.

4.4.2.2 The Formal Schema for Commitments

Definition 4.7 *A general schema covering different types of collective commitment is the following, where the conjuncts between curly brackets may be present or not; the slash (/) abbreviates a choice between two possibilities:*

$$\text{C-COMM}_{G,P}(\varphi) \leftrightarrow \tag{4.1}$$

$$\text{M-INT}_G(\varphi) \wedge \{awareness_G^1(\text{M-INT}_G(\varphi))\} \wedge$$

$$constitute(\varphi, P) \wedge \{awareness_G^2(constitute(\varphi, P))\} \wedge$$

$$\bigwedge_{\alpha \in P} \bigvee_{i,j \in G} \text{COMM}(i, j, \alpha) \wedge$$

$$\{awareness_G^3(\bigwedge_{\alpha \in P, j \in G} \bigvee \text{COMM}(i, j, \alpha)) \ / \ \bigwedge_{\alpha \in P, j \in G} \bigvee awareness_G^3(\text{COMM}(i, j, \alpha))\}$$

---

[2] Note that in this book we do not use predicate logic, so quantifiers like $\exists x$ are not used in our theory of teamwork. Instead of existential and universal quantifiers over a possibly infinite domain, we use finite disjunctions and conjunctions, as we deal with finite domains only.

Group G has a collective commitment to achieve overall goal φ based on social plan P (C-COMM$_{G,P}(\varphi)$) if and only if at least the following hold. The group mutually intends φ; moreover, there is the property $constitute(\varphi, P)$ implying that successful execution of social plan P leads to φ (see Section 6.4.1), and finally, for every action α from social plan P, there is one agent in the group who is socially committed to another group member to fulfil the action. Even though this does not often happen, self-commitments (where $i = j$) are allowed in this context.

Instantiating the above schema corresponds to tuning the $awareness_G^i$-dials from $\emptyset$, that is lack of any awareness, through individual beliefs BEL and different degrees of general belief E-BEL$_G^k$, to common belief C-BEL$_G$.

4.4.3 A Paradigmatic Group Commitment

The notion of collective commitment, whichever strength of it is considered, combines essentially different aspects of teamwork: strictly technical ones related to social plans, as well as those related to the agents' intentional stance. The degree of awareness is characterized in terms of different types of beliefs. To make the schema more concrete, a typical and relatively strong example of collective commitment is shown. The explanation makes a clear difference between the two types of awareness. Below, a strong type of awareness is considered, namely common belief: $awareness_G^1$, $awareness_G^2$ and $awareness_G^3$ are set to C-BEL$_G$ in the general schema of Definition 4.7. Therefore it is justified to speak about *collective awareness* in this context.

Let us discuss the relevant aspects in detail.

1. Collective intention (built on mutual intention and a common belief about this) is the attitude constituting the team as a whole:
   C-BEL$_G$(M-INT$_G(\varphi)$)
2. Collective awareness about the details of the plan P and its a priori correctness with respect to φ:
   C-BEL$_G$($constitute(\varphi, P)$)
3. Collective awareness about the distribution of actions and pairwise social commitments. In this way a team structure is created and the plan becomes social. The type of awareness connected with this phase may be twofold.
   (a) A *detailed* collective awareness of each social commitment:
       $\bigwedge_{\alpha \in P} \bigvee_{i,j \in G}$ C-BEL$_G$(COMM(i, j, α))
   (b) A *global* collective awareness of the bare existence of social commitments:
       C-BEL$_G$($\bigwedge_{\alpha \in P} \bigvee_{i,j \in G}$ COMM(i, j, α))

Note that C-BEL$_G$ in the detailed and global awareness distributes over conjunction ($\bigwedge_{\alpha \in P}$), so that only the position of C-BEL$_G$ with respect to $\bigvee_{i,j \in G}$ matters. We give a lemma about the relation between the two types of collective awareness:

Lemma 4.1 Detailed collective awareness ($\bigwedge_{\alpha \in P} \bigvee_{i,j \in G}$ C-BEL$_G$(COMM(i, j, α))) implies global collective awareness (C-BEL$_G$($\bigwedge_{\alpha \in P} \bigvee_{i,j \in G}$ COMM(i, j, α))), but not vice versa.

Proof We work in a system that includes $KD45_n^C$ for individual and common belief. Let us reason semantically.

Suppose that detailed awareness holds in a world $\mathcal{M}, s$:

$$\mathcal{M}, s \models \bigwedge_{\alpha \in P} \bigvee_{i,j \in G} \text{C-BEL}_G(\text{COMM}(i, j, \alpha)).$$

Now take any $\alpha \in P$, then there is a pair $i, j \in G$ such that:

$$\mathcal{M}, s \models \text{C-BEL}_G(\text{COMM}(i, j, \alpha)),$$

so a fortiori, by propositional logic and common belief distribution:

$$\mathcal{M}, s \models \text{C-BEL}_G(\bigvee_{i,j \in G} \text{COMM}(i, j, \alpha))$$

As $\alpha \in P$ was arbitrary, we may conclude that:

$$\mathcal{M}, s \models \bigwedge_{\alpha \in P} \text{C-BEL}_G(\bigvee_{i,j \in G} (\text{COMM}(i, j, \alpha)))$$

which is equivalent to:

$$\mathcal{M}, s \models \text{C-BEL}_G(\bigwedge_{\alpha \in P} \bigvee_{i,j \in G} (\text{COMM}(i, j, \alpha)))$$

because in general:

$$KD45_n^C \vdash \text{C-BEL}_G(\psi_1 \wedge \ldots \wedge \psi_n) \leftrightarrow \text{C-BEL}_G(\psi_1) \wedge \ldots \wedge \text{C-BEL}_G(\psi_n)$$

The converse does not hold. Take for example group $G = \{1, 2, 3\}$ and plan:

$$P = \langle \langle \langle 1, a \rangle; \langle 2, b \rangle \rangle; \langle 3, c \rangle \rangle$$

where each social commitment is commonly believed only by its two participants but not by the group as a whole:

$$\mathcal{M}, w \models \text{COMM}(1, 2, a) \wedge \text{COMM}(2, 3, b) \wedge \text{COMM}(3, 1, c) \wedge$$

$$\neg \text{C-BEL}_G(\text{COMM}(1, 2, a)) \wedge \neg \text{C-BEL}_G(\text{COMM}(2, 3, b)) \wedge$$

$$\neg \text{C-BEL}_G(\text{COMM}(3, 1, c))$$

Still, there is collective awareness that for each action, some social commitment is in place. In such a case, there is global awareness without detailed awareness about the three social commitments:

$$\mathcal{M}, w \models \text{C-BEL}_G(\bigwedge_{\alpha \in P} \bigvee_{i,j \in G} (\text{COMM}(i, j, \alpha)))$$

4.5   Different Notions of Collective Commitment

In order to make the theory of collective commitments more concrete, we will now instan-
tiate the general schema in five different ways. All of these lead to group commitments
actually occurring in the practice of different organization types.

The following exemplary definitions are formulated by keeping the *awareness*$_G^i$-dials
fixed to either $\emptyset$ or common belief C-BEL$_G$. We will start from the strongest possible
form of collective commitment, fully reflecting the collective aspects of teamwork. Subsequently, some of the underlying assumptions will be relaxed, leading ultimately to weaker
commitments.

4.5.1   Robust Collective Commitment

Robust collective commitment (R-COMM$_{G,P}$) is the strongest type of group commitment,
produced by instantiating all the awareness-dials in the general schema of Definition 4.7
to C-BEL$_G$, together with the detailed (or *de re*) collective awareness about bilateral
commitments.

$$R\text{-COMM}_{G,P}(\varphi) \leftrightarrow C\text{-INT}_G(\varphi) \wedge$$

$$constitute(\varphi, P) \wedge C\text{-BEL}_G(constitute(\varphi, P)) \wedge$$

$$\bigwedge_{\alpha \in P} \bigvee_{i,j \in G} COMM(i, j, \alpha) \wedge$$

$$\bigwedge_{\alpha \in P} \bigvee_{i,j \in G} C\text{-BEL}_G(COMM(i, j, \alpha))$$

Intuitively, robust collective commitment may be based on collective planning, including negotiating and persuading one another who will do what. In effect, for every action α
from social plan P, one agent i has socially committed to another team member j to fulfill
the action α: COMM(i, j, α). Moreover, the team as a whole is aware of every single
social commitment that has been established. Thus the social structure of the team as well
as everybody's responsibility is public. The aspect of sharing responsibility is essential
here. Among others it implies that there is no need for an initiator in such a team.

Example 4.1 *Robust collective commitment may be applicable in (small) companies where
all team members involved are share-holders. Typically, planning is done collectively.
Everybody's responsibility is public because the social commitments are generally known.
In particular, when any form of revision is needed due to dynamic circumstances, the entire
team may be collectively involved.*

*This type of collective commitment is most suited for self-leading teams, which are not
directly led by a manager. Instead, the team is responsible for achieving some high-level
goals, and is entirely free to divide roles, devise a plan, etc. (Beyerlin et al., 1994; Purser
and Cabana, 1998). A non-hierarchical team of researchers is a typical example of such
a self-leading team establishing a robust collective commitment.*

4.5.2   Strong Collective Commitment

As in the case of robust collective commitment, when instantiating the general schema for *strong collective commitment* (S-COMM$_{G,P}$), all *awareness$_G$*-dials are placed at C-BEL$_G$. However, as to awareness of social commitments, global (or de dicto) collective awareness is applied, making strong collective commitment somewhat weaker than the robust one:

$$\text{S-COMM}_{G,P}(\varphi) \leftrightarrow \text{C-INT}_G(\varphi) \wedge$$

$$constitute(\varphi, P) \wedge \text{C-BEL}_G(constitute(\varphi, P)) \wedge$$

$$\bigwedge_{\alpha \in P} \bigvee_{i,j \in G} \text{COMM}(i, j, \alpha) \wedge$$

$$\text{C-BEL}_G \Big(\bigwedge_{\alpha \in P} \bigvee_{i,j \in G} \text{COMM}(i, j, \alpha) \Big)$$

Intuitively, in contrast to R-COMM$_{G,P}$, in strong collective commitment, there is no detailed public awareness about particular social commitments, but the group as a whole believes that things are under control, that is that every part of the plan is within somebody's responsibility. This implies that the social structure of the team is not public. As the responsibility is not shared, the case of a team leader or initiator fits here. Also, as pairwise social commitments are not publicly known, they cannot be collectively revised when such a need appears.

Example 4.2 *Strong collective commitment may be applicable in companies with one or more leaders and rather separate subteams. Even though planning may be done collectively, establishing bilateral commitments is not done publicly, but in subgroups. For example, members might promise their sub-team leader to do their own part.*

In many situations this global awareness of social commitments suffices and is preferred for efficiency reasons.

4.5.3   Weak Collective Commitment

In the somewhat less demanding *weak collective commitment* (W-COMM$_{G,P}$), the degree of team awareness is more limited. Formally, weak commitments are distinguished from strong ones by instantiating the *awareness$_G^2$*-dial at $\emptyset$:

$$\text{W-COMM}_{G,P}(\varphi) \leftrightarrow \text{C-INT}_G(\varphi) \wedge constitute(\varphi, P) \wedge$$

$$\bigwedge_{\alpha \in P} \bigvee_{i,j \in G} \text{COMM}(i, j, \alpha) \wedge$$

$$\text{C-BEL}_G \Big(\bigwedge_{\alpha \in P} \bigvee_{i,j \in G} \text{COMM}(i, j, \alpha) \Big)$$

As usual, the team knows the overall goal but does not know details of the plan: there is no collective awareness of the plan's correctness. Apparently, also in this case no collective revision of social commitments may take place.

Example 4.3 *Weak collective commitment may be applicable in companies with a planning department or a dedicated planner, who individually believes in the plan's correctness and this should suffice. Paradigmatic examples of such companies are large multi-nationals with extensive planning departments.*

4.5.3.1 Remark about Robust, Strong and Weak Commitments

In the weak collective commitment, the team does not know the plan details and cannot have any opinion about its correctness. Still team members believe that things are under control and remain aware about their share in the plan. In the strong and robust case, the team additionally knows the plan details and is prepared to act accordingly.

Apparently, there is a plethora of other possibilities of group involvement in CPS, two of which are shown below: team commitment and distributed commitment. It often happens that agents' limited orientation in the labor division is done on purpose, even though the overall goal is known to everybody, due to the definition of collective intention.

4.5.4 Team Commitment

In team commitment (T-COMM$_{G,P}$), the *awareness*^1_G-dial is set to C-BEL$_G$, while both others are reduced to Ø:

$$\text{T-COMM}_{G,P}(\varphi) \leftrightarrow \text{C-INT}_G(\varphi) \wedge constitute(\varphi, P) \wedge$$

$$\bigwedge_{\alpha \in P} \bigvee_{i,j \in G} \text{COMM}(i, j, \alpha)$$

The presence of collective intention ensures that the team as a structure still exists and the overall goal and composition of the team are commonly believed. Planning is not at all collective, and it may be that even task division is not public: agents remain aware solely about their piece of work, without any orientation about involvement of others. Again, this is often done deliberately. Thus, distribution of social commitments cannot be public either.

Example 4.4 *Team commitment may be applicable in companies assigning limited trust to their employees. Information about the task allocation may be confidential.*

4.5.5 Distributed Commitment

The last case is *distributed commitment* (D-COMM$_{G,P}$) where all *awareness*$_G$-dials are set to Ø, and goes beyond the schema in not being based on a mutual intention:

$$\text{D-COMM}_{G,P}(\varphi) \leftrightarrow constitute(\varphi, P) \wedge \bigwedge_{\alpha \in P} \bigvee_{i,j \in G} \text{COMM}(i, j, \alpha)$$

This deals with the situation when agents' awareness is even more restricted than in team commitment: they do not even know the overall goal, solely their share in an 'undefined' project. As there is no collective intention C-INT$_G$, no 'real' team is created. Instead, a rather loosely coupled group of agents works in a distributed manner without autonomous involvement in the project.

Example 4.5 *Distributed commitment may be applicable in companies contracting out some labour to outsiders. The overall goal, the agents involved, as well as some other aspects of the project may be classified information, for example in order to avoid competition.*

Another typical case of distributed commitment is displayed by groups of 'spies' as introduced in the beginning of this chapter. In their case, lack of information about the tasks or even the identity of other group members may be beneficial to everybody's safety. They work with an inflexible plan set in advance by one 'mastermind', so their autonomy and flexibility are severely curtailed. As to efficiency, the need for communication when preparing and executing team action is limited as well.

4.5.6 Awareness of Group Commitment

The weaker notions such as team and distributed commitment are especially suited to model hierarchically organized teams, where power relations between team members are made explicit. The simplest case of agents' organization is teamwork completely controlled by an initiator. Though we refrain from introducing the power aspect explicitly in the definitions, their different strengths may be useful in various situations (see Castelfranchi *et al.*, 1992), especially when maintaining a balance between the centralized power and the spread of knowledge.

The stronger notions like robust and strong collective commitment are well-suited to model so-called self-leading teams which are currently studied in the organizational science literature (Beyerlin *et al.*, 1994). All agents involved are collectively aware of the situation:

theorem: awareness of robust collective commitment

$$\text{R-COMM}_{G,P}(\varphi) \rightarrow \text{C-BEL}_G(\text{R-COMM}_{G,P}(\varphi))$$

theorem: awareness of strong collective commitment

$$\text{S-COMM}_{G,P}(\varphi) \rightarrow \text{C-BEL}_G(\text{S-COMM}_{G,P}(\varphi))$$

The proofs are immediate from the definitions and Lemma 2.1.

Note that the theorem does not hold for weaker forms of group commitments, where agents' awareness is very limited. Therefore, they are not aware of the exact strength of the group commitment among them. More importantly, some intermediate levels of commitment may be characterized by replacing C-BEL$_G$ by E-BEL$_G$ when it suffices for the proper organization of teamwork.

4.6 Topologies and Group Commitments

The definitions of group commitments in Section 4.5 enable the system developer to organize teams or larger organizational structures according to a specific chosen type of commitment. However, in real applications much more complex distributed structures can be considered: sub-teams of agents, created on the basis of various commitments, may be combined into larger structures. Thus, heterogeneity of these structures is achieved. In order to cover the variety of possibilities, potential 'ties' in these complex organizations may be implemented in many different ways. One of them would be introducing an organization's social structure *explicitly*, for example by a labeled tree, in contrast to the *implicit* form adopted above. An explicit social structure has many advantages, as it gives an opportunity to appropriately organize various substructures within a complex framework. Thus, *scalability* of these organizations comes to the fore, a phenomenon that barely received prior attention. As indicated, when using such an explicit framework, specification of truly large organizations is possible, making them easier to predict.

The social structure of a group may originate from different types of topologies, based on power and dependency relations (Castelfranchi *et al.*, 1992; Dignum and Dignum, 2007; Gasser, 1995; Grossi, *et al.*, 2007). These relations are implicitly reflected in the setting of social commitments. For example, in a hierarchical tree structure, social commitments are made to the agent that is the direct ancestor in the tree. In some cases, it is worthwhile to make the team's implicit structure more explicit by adapting the schema according to the topology.[3]

4.6.1 Robust Commitments with a Single Initiator under Infallible Communication

Suppose that teamwork starts with a single initiator who tries to establish a robust social commitment. Suppose that a strictly cooperative team already exists and a collective intention is in place, following the standard axiom **M3** from Section 3.5:

$$\textbf{M3} \quad \text{C-INT}_G(\varphi) \leftrightarrow \text{M-INT}_G(\varphi) \wedge \text{C-BEL}_G(\text{M-INT}_G(\varphi))$$

Planning is done by the team as a whole. We assume that carrying out the plan P leads to achieving the team's goal φ. Every agent knows the plan and is convinced of its correctness. Moreover, let us assume as an idealization that the agents involved communicate through an infallible medium and that this is commonly believed. Thus, they broadcast their messages in such a way that every other agent receives them and is aware that the others did as well. Assuming that every agent reasons in the same way, a broadcast about the plan's correctness leads to:

$$\text{C-BEL}_G(constitute(\varphi, P))$$

During action allocation, agents who can carry out certain actions submit their promises to the *initiator*, again by public broadcasts. Therefore, every agent believes every declared promise and believes that other agents believe this as well. Thus, whenever agent a volunteers to perform α, a bilateral commitment between him and the *initiator* is formed:

$$\text{COMM}(a, initiator, \alpha)$$

The infallible broadcast results in a common belief about this fact:

$$\text{C-BEL}_G(\text{COMM}(a, initiator, \alpha))$$

---

[3] The description of examples in the subsequent subsections has been inspired by the work of Barbara's students at Warsaw University, namely Filip Grządkowski, Michał Modzelewski, Alina Strachocka and Joanna Zych.

Finally, the successful process of forming bilateral commitments results in:

$$\bigwedge_{\alpha \in P} \bigvee_{a \neq initiator} \text{COMM}(a, initiator, \alpha) \wedge$$

$$\text{C-BEL}_G (\bigwedge_{\alpha \in P} \bigvee_{a \in G \setminus \{initiator\}} \text{COMM}(a, initiator, \alpha))$$

To sum up, we have:

$$\text{C-INT}_G(\varphi) \wedge constitute(\varphi, P) \wedge$$

$$\text{C-BEL}_G(constitute(\varphi, P)) \wedge \bigwedge_{\alpha \in P} \bigvee_{a \in G - \{initiator\}} \text{COMM}(a, initiator, \alpha) \wedge$$

$$\text{C-BEL}_G (\bigwedge_{\alpha \in P} \bigvee_{a \in G \setminus \{initiator\}} \text{COMM}(a, initiator, \alpha))$$

This clearly implies that there is a robust commitment $\text{R-COMM}_{G, P}(\varphi)$ among the team G and one can easily read off the specific structure of the team from the social commitments towards the single initiator.

4.6.2 Star Topology with a Single Initiator under Restricted Communication

Let us assume that among a strictly cooperative team of agents there exists a dedicated *initiator*, who creates a plan P, leading to the goal φ, so that $constitute(\varphi, P)$ holds. Subsequently, the information about the plan is passed to group members individually. Each message sent to agent a_i is received only by this agent. We have therefore: $\text{E-BEL}_G(constitute(\varphi, P))$, but not $\text{C-BEL}_G(constitute(\varphi, P))$.

The *initiator* knows the individual capabilities of team members and allocates tasks accordingly: for each action α_i he assigns an agent a_i to execute it. After successful action allocation, the agents are individually informed about the results. In this way bilateral commitments emerge between the *initiator* and other agents:

$$\bigwedge_{\alpha \in P} \bigvee_{a \in G \setminus \{initiator\}} \text{COMM}(a, initiator, \alpha)$$

The second part of the initiator's message to each individual agent leads to:

$$\text{E-BEL}_G (\bigwedge_{\alpha \in P} \bigvee_{a \in G \setminus \{initiator\}} \text{COMM}(a, initiator, \alpha))$$

Yet none of the agents except for *initiator* knows anything about the state of others' beliefs. In this case a collective commitment comes to:

$$\text{C-INT}_G(\varphi) \wedge constitute(\varphi, P) \wedge$$

$$\text{E-BEL}_G(constitute(\varphi, P)) \wedge \bigwedge_{\alpha \in P} \bigvee_{a \in G \setminus \{initiator\}} \text{COMM}(a, initiator, \alpha) \wedge$$

$$\text{E-BEL}_G (\bigwedge_{\alpha \in P} \bigvee_{a \in G \setminus \{initiator\}} \text{COMM}(a, initiator, \alpha))$$

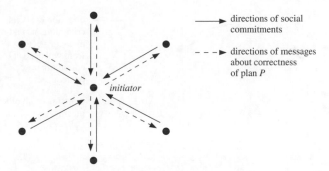

Figure 4.1 Team structure of the star topology illustrating the messages and social commitments. The *initiator* first sends individual messages '*constitute*(φ, P)' to all team members and after each $a \in G$ pronounces its social commitment COMM(a, *initiator*, α), the *initiator* sends each team member individually the message '$\bigwedge_{\alpha \in P} \bigvee_{a \in G \setminus \{initiator\}}$ COMM(a, *initiator*, α)'.

This type of commitment is weaker than the weak collective commitment, due to the presence of merely general beliefs instead of common ones. Still this may suffice in some applications, as team members clearly have more information than in the case of team commitment. Again, it is easy to read off the team's star topology from the social commitments (see Figure 4.1).

4.6.3 Ring Topology with a Single Initiator

Let us define a type of team planning inspired by the children's game *Chinese whispers* (see Figure 4.2). In this game, an initiator thinks of a message and whispers it to the next person. All other group members must carry out their pre-defined task of passing the message along until it travels all the way around the circle. When trying the game in practice it turns out that this type of communication is prone to errors caused by malicious or broken agents in the communication path.

This game can be adapted to a type of teamwork where an initiator is responsible for *planning* and specific tasks are passed around a communication ring. Each member of the group picks a number of tasks to realise and passes the remaining tasks along. Thus, assume that the group of agents G communicates according to a ring topology. The dedicated agent a_0 (initiator) composes a sequence of actions τ which together are sufficient to achieve the goal φ, without any assumptions on their temporal order. This allows the condition *constitute*(φ, P) to hold for any plan P based on τ. After composing the sequence, the initiator sends a message consisting of the sequence $\tau = \alpha_1, \ldots, \alpha_n$ and a vector $\vec{v}_0 = \vec{0}$ of size n.

Each agent a_j (for $j \geq 1$), when receiving the message containing the details of the sequence τ and the vector $\vec{v}_{j-1}$, can verify its correctness and deduce *constitute*(φ, P) for all plans based on τ. Let us abbreviate by *leadsto*(τ, φ) the fact that *constitute*(φ, P) holds for all plans based on τ. Moreover, a_j chooses a set of actions $\tau_{a_j} \subseteq \tau$ for which $\bigwedge_{\alpha_i \in \tau_{a_j}} \vec{v}_{j-1}(i) = 0$, that it can then carry it out. Then the agent creates a vector $\vec{v}_j$ such that:

$$\bigwedge_{\alpha_i \in \tau_{a_j}} \vec{v}_j(i) = 1$$

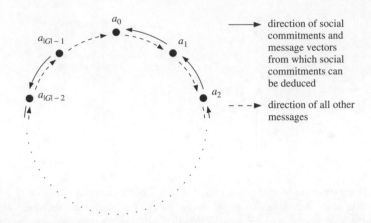

Figure 4.2 Team structure of the ring topology illustrating the messages and social commitments in the 'Chinese whispers procedure', starting from a message by *initiator* to a_1. For $j \geq 1$, each a_j receives a message from a_{j-1} containing action sequence τ and a vector $\vec{v}_{j-1}$ representing previously chosen actions. Then a_j chooses its own actions, commits to do them with respect to a_{j-1} and then sends on τ and the current allocated actions vector $\vec{v}_j$ to a_{j+1}. In the second half of the procedure, a_0 passes around the message ψ meaning that all actions have been allocated and that everyone believes the plan to be correct. This may be repeated, where in the kth round the agents send around the message ψ_k, abbreviating $\psi \wedge \text{E-BEL}_G^k(\psi)$, finally leading after M iterations to an M-fold general belief that the plan is fine and they're all set for team action.

and:

$$\bigwedge_{\alpha_i \notin \tau_{a_j}} \vec{v}_j(i) = \vec{v}_{j-1}(i)$$

and sends it to agent a_{j-1}. This is regarded as a commitment to execute the set of actions τ_{a_j} it has chosen:

$$\bigwedge_{\alpha \in \tau_{a_j}} \text{COMM}(a_j, a_{j-1}, \alpha)$$

Eventually, it forwards the sequence τ and the vector $\vec{v}_j$ to agent a_{j+1} and receives from it a message about a_{j+1}'s chosen actions (also a vector $\vec{v}_{j+1}$). When agent a_0 receives back a vector composed of the previous contributions of ring members it can deduce the following proposition, that we abbreviate by ψ:

$$\bigwedge_{\alpha \in \tau} \bigvee_{j \in \{1, \ldots, |G|-1\}} \text{COMM}(a_j, a_{j-1}, \alpha) \wedge$$

$$\text{E-BEL}_G(\textit{leadsto}(\tau, \varphi))$$

Then, initiator a_0 sends the above proposition ψ as a message to the next agent in the ring. As soon as it receives ψ back from the last agent $a_{|G-1|}$, $\text{BEL}(a_0, \psi \wedge \text{E-BEL}_G(\psi))$ starts to hold. This fact can also be communicated within the ring, and so forth, where in each round k the message $\psi_k \equiv \psi \wedge \text{E-BEL}_G^k(\psi)$ is sent around the ring. Depending on

M, the number of rounds in the ring, we will have:

$$\text{E-BEL}_G^M \left(\bigwedge_{\alpha \in \tau} \bigvee_{j \in \{1,\dots,|G|-1\}} \text{COMM}(a_j, a_{j-1}, \alpha) \right) \wedge$$

$$\text{E-BEL}_G^{M+1}(leadsto(\tau, \varphi))$$

Thus, the collective commitment in team G based on sequence τ will be as follows:

$$\text{C-INT}_G(\varphi) \wedge leadsto(\tau, \varphi) \wedge \text{E-BEL}_G^{M+1}(leadsto(\tau, \varphi)) \wedge$$

$$\bigwedge_{\alpha \in \tau} \bigvee_{j \in \{1,\dots,|G|-1\}} \text{COMM}(a_j, a_{j-1}, \alpha) \wedge$$

$$\text{E-BEL}_G^M \left(\bigwedge_{\alpha \in \tau} \bigvee_{j \in \{1,\dots,|G|-1\}} \text{COMM}(a_j, a_{j-1}, \alpha) \right)$$

In conclusion, the group has achieved an attitude which is a bit weaker than strong collective commitment $\text{S-COMM}_{G,P}(\varphi)$, due to the presence of an M-fold general belief instead of common belief. For many practical purposes, such iterated general beliefs are sufficient to support teamwork. The required number of iterations depends on the specific situation. At design time, the trade-off between a team's certainty about motivational attitudes and time spent on communication needs to be balanced.

4.6.4 A Hierarchical Group: Trees of Shallow Depth

Suppose that a certain group G has as its leader a *planner* and that *subleader*$_1, \dots,$ *subleader*$_n$ lead their own subgroups $G_1, \dots, G_n$ and are intermediaries in any communication between the *planner* and the subgroups. Let us suppose that the collective intention $\text{C-INT}_G(\varphi)$ is already present. The leader has made a correct overall plan P, consisting of n subplans $P_1, \dots, P_n$ that correspond to the n subgoals $\varphi_1, \dots, \varphi_n$ of φ. The leader believes that plan P indeed leads to achievement of φ, so we have $constitute(\varphi, P) \wedge \text{BEL}(planner, constitute(\varphi, P))$. This includes the leader's belief in the correctness of each subplan:

$$\bigwedge_{i=1}^n \text{BEL}(planner, constitute(\varphi_i, P_i))$$

When delegating these subplans to the subleaders, the main *planner* communicates privately with each of them about the correctness of their delegated subplan. This results in the subleaders' beliefs in the correctness of the parts for which they are responsible:

$$\bigwedge_{i=1}^n \text{BEL}(subleader_i, constitute(\varphi_i, P_i))$$

Each *subleader*$_i$ delegates actions from subplan P_i to members of its subteam G_i, and the responsible agents commit to perform them:

$$\bigwedge_{i=1}^n \bigwedge_{\alpha \in P_n} \bigvee_{j \in G_n} \text{COMM}(j, subleader_i, \alpha)$$

The subleader then communicates to the *planner* that all actions have been delegated, but without giving the exact details, thus resulting in the leader's global belief that everything is under control:

$$\bigwedge_{i=1}^{n} \text{BEL}(planner, \bigwedge_{\alpha \in P_n} \bigvee_{j \in G_n} \text{COMM}(j, subleader_i, \alpha))$$

Altogether, the team's collective commitment amounts to:

$$\text{C-COMM}_{G,P}(\varphi) \leftrightarrow \text{C-INT}_G(\varphi) \wedge$$

$$constitute(\varphi, P) \wedge \bigwedge_{i=1}^{n} constitute(\varphi_i, P_i) \wedge$$

$$\text{BEL}(planner, constitute(\varphi, P)) \wedge \bigwedge_{i=1}^{n} \text{BEL}(planner, constitute(\varphi_i, P_i)) \wedge$$

$$\bigwedge_{i=1}^{n} \text{BEL}(subleader_i, constitute(\varphi_i, P_i)) \wedge$$

$$\bigwedge_{i=1}^{n} \bigwedge_{\alpha \in P_i} \bigvee_{j \in G_i} \text{COMM}(j, subleader_i, \alpha) \wedge$$

$$\bigwedge_{i=1}^{n} \text{BEL}(planner, \bigwedge_{\alpha \in P_i} \bigvee_{j \in G_i} \text{COMM}(j, subleader_i, \alpha))$$

As in the case of the star topology with limited communication, this implies a collective commitment stronger than a team commitment $\text{T-COMM}_{G,P}(\varphi)$, due to the extra information that the leader and subleaders have about the social commitments. On the other hand, it is not sufficient for a weak social commitment $\text{S-COMM}_{G,P}(\varphi)$, because the leader is the only team member with access to the information that all tasks have been properly assigned. As with the other topologies, the specific tree structure can be easily read off the social commitments towards the subleaders and the main leader, the *planner* (see Figure 4.3).

4.7   Summing up TEAMLOG: The Static Part of the Story

We have incrementally built TEAMLOG, a static theory of teamwork, starting from individual intentions, through social commitments, leading ultimately to collective intentions and collective commitments. These notions are defined in multi-modal logics with clear semantics, comprising a descriptive view on teams' motivational attitudes. While developing our ideas presented for the first time in Dunin-Kęplicz and Verbrugge (1996), we became more and more flexible about adjusting the notion of collective commitment to current circumstances, instead of aiming for one 'iron-clad' reading of group commitment. Therefore, we provide a sort of tuning mechanism for the system developer to calibrate an appropriate type of group commitment, taking into account both the circumstances in

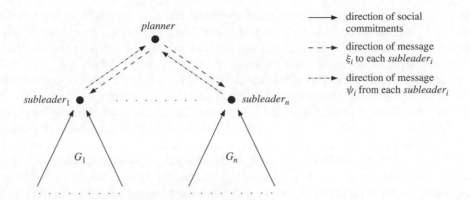

Figure 4.3   Team structure of the shallow tree topology illustrating the messages and social commitments. The *leader* sends each *subleader$_i$* the message ξ_i about the correctness of delegated subplans: '*constitute*(φ_i, P_i)'. After agents from the subteam G_i have committed to all its actions, *subleader$_i$* sends its final answer ψ_i, abbreviating '$\bigwedge_{\alpha \in P_i} \bigvee_{j \in G_i} \text{COMM}(j, subleader_i, \alpha)$' back to the *planner*.

which a group is acting (like communication capabilities), as well as the possibly complex organizational structure including power and dependency relations. The multi-modal logic framework allows us to express subtle aspects of teamwork while modeling different situations occurring in real applications.

4.7.1   Comparison

The characteristics of various group commitments in Section 4.5 are not overloaded and therefore easy to understand and to use. Some other approaches to collective and joint commitments (see for example Levesque *et al.*, 1990 and Wooldridge and Jenning, (1999)) introduce other aspects, not treated here. For example, Wooldridge and Jennings (1999) consider triggers for commitment adoption formulated as preconditions. As another example, Aldewereld *et al.* (2004) add constraints about goal adoption and achievement to their definitions of joint motivational attitudes. As indicated previously, we have chosen to incorporate solely vital aspects of the defined attitudes, leaving room for any case-specific extensions. If needed, these extensions may be incorporated by adding extra axioms. Note that in contrast to other approaches such as Levesque *et al.* (1990) and Wooldridge and Jennings (1999), the collective commitment is not 'iron-clad': it may vary in order to adapt to changing circumstances, in such a way that the collective intention on which it is based can still be reached.

4.7.2   Moving Towards a Dynamic View on Teamwork

In this chapter, we do not describe how collective intentions, and then collective commitments, are actually established in a group. Intention and commitment adoption will be treated in the coming Chapters 5, 6 and 8 (see also Dignum *et al.*, 2001b and Dunin-Kęplicz and Verbrugge, 2001b). The dynamic part of the story of teamwork, TEAMLOG^{dyn}, will be introduced in the next two chapters.

In fact, our approach to collective motivational attitudes is especially strong when re-planning is needed. In contrast to Wooldridge and Jennings (1999), using our notions of collective commitment it is often sufficient to revise some of the pairwise social commitments, instead of involving the entire team in re-planning, particularly in the strong types of collective commitments. This efficiency is entailed by building collective commitments from an explicit plan representation and bilateral social commitments. In effect, if the new plan resulting from the analysis of the current circumstances is as close as possible to the original one, re-planning is nearly optimal. This reconfiguration problem is presented extensively in Chapter 5. Subsequently, in Chapter 6, the reconfiguration procedure is proved to be correct. Finally, in Chapter 8, the dialogues among computational agents involved in teamwork are made transparent. Thus, the next three chapters contribute to the *dynamic*, more prescriptive theory of collective motivational attitudes. Combining the static theory TEAMLOG with dynamic aspects, the full theory TEAMLOG^{dyn} may serve a system designer as a specification of a correct system.

5

Reconfiguration in a Dynamic Environment

Do you have the patience to wait
till your mud settled and the water is clear?
Can you remain unmoving
till the right action arises by itself?

Tao Te Ching (Lao-Tzu, Verse 15)

5.1 Dealing with Dynamics

A dynamic environment requires flexible behavior to ensure successful teamwork. Even though the first stages of teamwork have been extensively discussed in the MAS and AI literature (Cohen *et al.*, 1997; Nair *et al.*, 2003; Pynadath and Tambe, 2002; Shehory, 2004; Shehory and Kraus, 1998), the resulting team action (or plan execution) has received relatively little attention. Let us analyze this phase now.

To maintain a collective intention during plan execution, it is vital that agents replan properly and efficiently in accordance with the circumstances. When some team members cannot realize their individual actions, or, on the positive side, some others are presented with new opportunities, re-planning takes place. This intelligent re-planning is the essence of the *reconfiguration problem*, discussed for the first time independently by Tambe (1996, 1997) and by Dunin-Kęplicz and Verbrugge (1996, 2001a).

During reconfiguration, adaptations of the original plan may be done from scratch, for the price of losing what has been achieved before. For resource-bounded agents, it is much more efficient to smartly adapt the previous results to the current situation. Such intelligent re-planning implies a natural *evolution* of the team's commitment, including the evolution of plans and motivational attitudes involved. These changes are methodologically treated in a generic *reconfiguration algorithm* formulated in terms of the consecutive stages of teamwork and their complex interplay.

Teamwork in Multi-Agent Systems: A Formal Approach   Barbara Dunin-Kęplicz and Rineke Verbrugge
© 2010 John Wiley & Sons, Ltd

5.1.1   Collective Commitments in Changing Circumstances

Now that all teamwork attitudes have been intuitively and formally characterized in the static part of TEAMLOG (see Chapters 2, 3 and 4), they should be confronted with the paradigmatic situation justifying their creation, namely a complex, dynamic environment. After all, the proof of the pudding is in the eating.

To study teamwork and its dynamics, we will isolate and analyse separately the three essential aspects of team cooperation and coordination in a distributed environment. These are *construction*, *maintenance* and *realization* of the type of collective commitments that optimally fit the application domain, the group structure and the situation. This can be done by the system developer at design time or by the initiator at runtime. On the general issue of tuning collective commitments, see Chapter 4. Throughout this chapter, we will use the generic notion of collective commitment C-COMM$_{G,P}$, abstracting from any particular type of commitment. As reconfiguration amounts to intelligent replanning, we will naturally focus on a team's *social plan* which is an obligatory element of *any* group commitment. This way our approach to reconfiguration acquires universality, transcending commitment types.

5.1.2   Three Steps that Lead to Team Action

In many BDI systems, teamwork is modeled explicitly (Aldewereld *et al.*, 2004; Grosz and Kraus, 1996; Levesque *et al.*, 1990; Tambe, 1996, 1997; Wooldridge and Jennings, 1999). An explicit model helps the team to monitor its performance and to re-plan efficiently, in accordance with the circumstances, when team members fail to realize their actions or new opportunities appear. A commonly recognized model of cooperative problem solving (CPS) has been provided by Wooldridge and Jennings (1996, 1999). We adapted their four-stage model, containing the consecutive stages of *potential recognition*, *team formation*, *plan formation* and *team action* for the sake of our analysis. However, especially with respect to collective intentions and collective commitments, our approach differs from the one in Wooldridge and Jennings (1996, 1999).

As advocated above, we study teamwork starting from potential recognition, assuming for simplicity that there is an agent-initiator who knows the overall goal φ and takes the initiative to realize it. This initiator is responsible for potential for cooperation among agents available at the time. The next step is team formation, leading to a collective intention between members of a successfully created team. The subsequent stage of planning, realized collectively in the most advanced case, results in the strongest motivational attitude, that is collective commitment. These complex preparations are finally concluded in team action.

An unpredictable and dynamic environment strongly influences teamwork, which becomes unpredictable to some extent when adjusting to actual circumstances. Therefore, modeling teamwork requires methods and techniques reflecting dynamics of its stages. Most of the time these methods originate from (Distributed) Artificial Intelligence; however, their specific variants have been created especially for multi-agent applications; see Durfee (2008), Jennings and Wooldridge (2000) and Wooldridge (2009) for extensive discussions.

In our reconfiguration story, we abstract from strictly technical aspects like methods and algorithms meant to realize stage-related procedures. Instead, our primary methodological goal is to characterize the stages of teamwork in a generic way, with an emphasis on their cooperative essence: the evolution of informational and motivational attitudes of team members. We focus on defining the final results of these stages in terms of agents' motivational stance. Such an approach will be profitable in clarifying the nature of dependencies between the agents involved. For example, some of them do domain problem solving, while others are responsible for the proper organization of teamwork.

The rest of this chapter is structured as follows. Section 5.2 presents a detailed introduction to the four stages of teamwork, including formal definitions of corresponding agent attitudes. Subsequently, the general ideas behind our reconfiguration method are explained and the reconfiguration algorithm is presented in the central Section 5.3. Finally, the algorithm is illustrated by an example application, extensively discussed in Section 5.4.

5.2   The Four Stages of Teamwork

For simplicity, we assume that the main goal of teamwork has been fixed. The important topic of goal selection is beyond the scope of this chapter and has been extensively treated elsewhere (Dignum and Conte, 1997; van der Hoek *et al.*, 2007).

5.2.1   Potential Recognition

Analogous to Wooldridge and Jennings (1999), we consider teamwork to begin when the initiator in a multi-agent environment recognizes the potential for cooperative action in order to reach its goal. The stage of potential recognition concerns finding the set of agents that may participate in the formation of the team aiming to realize the common goal. Potential team members should not only be willing to cooperate with others but they should also have relevant skills and resources at their disposal. Hence, potential recognition is a complex process, leading ultimately to a (hopefully non-empty) collection of potential teams with whom further discussion will follow during *team formation*.

The *input* of potential recognition is an initiator a, a goal φ plus a finite set $T \subseteq A$ of agents, potential team members. The successful outcome is the 'potential for cooperation' that the initiator a sees with respect to φ, denoted by the predicate PotCoop(φ, a). The first task of the *initiator* is to form a partial plan for achieving the main goal.[1] According to distinguished subgoals, it will determine characteristics of agents most suited to form the team. In order to determine this match, the initiator needs to find out relevant properties of the agents, being interested in four aspects: their *abilities*, *opportunities*, *willingness* to work together and individual *type*. An agent's *ability* concerns its subjective skills to perform the right type of action, regardless of the situation. This inherent property of the agent is contrasted with its *opportunity*, referring to resources and other application-related properties, related to the present state of the environment. Thus, opportunities form the objective view of agents' possibilities. For an in-depth discussion and formalization of abilities and opportunities, see van Linder *et al.* (1998). Next, *willingness* expresses

---

[1] See the literature on partial global planning and continual distributed planning (desJardins *et al.*, 1999; Durfee, 2008).

the agent's mental attitudes towards participating in team action. Very capable agents that do not want to do the job or are too busy to join the team are of no use.

It is needless to say that agents are diverse. As individual *type* of an agent we distinguish between software agents or artifacts like robots, unmanned aerial vehicles (UAVs) or, finally, human beings. After all, the initiator needs to compose an appropriate type distribution to make a future team effective. In this context, commitment strategies of potential team members, reflecting their 'characters', play a role (see Section 4.3.4). The combination of relevant properties is summarized in the predicate *propertypedistr* (G, φ), standing for 'group G has a proper distribution of agent types for achieving φ'.

5.2.1.1 Gathering Proper Agents for Possible Teams

The components of the agents' suitability are listed below; they can also be represented with respect to actions (like α) instead of the states of affairs (such as φ and ψ) appearing below:

1. An agent b's ability to achieve a goal ψ is denoted by $able(b, \psi)$.
2. An agent b's opportunity to achieve ψ is denoted by $opp(b, \psi)$.
3. The combination of an individual agent b's ability and opportunity to achieve ψ is summarized in $can(b, \psi)$:

$$can(b, \psi) \leftrightarrow able(b, \psi) \land opp(b, \psi).$$

   (See van Linder *et al.* (1998) for a formal treatment.)
4. A collective possibility to achieve φ by group G is denoted by $c\text{-}can_G(\varphi)$.
   (See Wooldridge and Jennings (1999) and van Linder *et al.* (1998) for formalizations of joint ability and 'collective can'.)
5. The willingness of agent b to participate in team formation is denoted by $willing(b, \varphi)$.
6. Whether there is a proper type distribution of agents in the considered group G to achieve φ is denoted by the predicate *propertypedistr* (G, φ). For an informal discussion of appropriate commitment strategy type distributions for several areas of application, see our paper (Dunin-Kęplicz and Verbrugge, 1996).

Now, we are ready to define a 'potential for cooperation' PotCoop(φ, a) that the initiator a sees with respect to φ. It includes:

- φ is a goal of a (GOAL(a, φ));
- there is a group G such that a believes that G can collectively achieve φ ($c\text{-}can_G(\varphi)$) and that the members of G are willing to participate in team formation;
- either a cannot or does not aim to achieve φ in isolation:

$$\text{PotCoop}(\varphi, a) \leftrightarrow \text{GOAL}(a, \varphi) \land$$

$$\exists G \subseteq T \ (\text{BEL}(a, c\text{-}can_G(\varphi) \land \bigwedge_{i \in G} willing(i, \varphi) \land$$

$$propertypedistr(G, \varphi))) \land$$

$$(\neg can(a, \varphi) \lor \neg\text{GOAL}(a, done(\texttt{stit}(a, \varphi))))).$$

Clearly there may be many groups G for which the initiator a finds out the right combination of initial properties $(c\text{-}can_G(\varphi) \wedge \bigwedge_{i \in G} willing(i, \varphi) \wedge propertypedistr(G, \varphi))$. Actually, this surplus may turn out to be very useful in the future course of action: if the current potential team does not work out, the initiator can pick the next available group from the collection of potential teams $H = (G_1, \ldots, G_n)$. Note that H may be of size exponential in the size of T; in practice, however, there is no need to store them all. Given collection H, the quantification $\exists G \subseteq T$ in the above formula may be replaced by $\exists G \in H$.

Starting from these initial properties, the adoption of the necessary motivational attitudes by the agents will be realized during team formation and plan formation.

5.2.2   Team Formation

Suppose that initiator a sees the potential for cooperation to achieve φ. Somewhat different from Wooldridge and Jennings (1999), we find that during the next stage of team formation agent a attempts to establish a *collective intention* towards φ in some group G. In our terminology this means that this group becomes a strictly cooperative team towards the goal φ.

The input of team formation is initiator a, goal φ and collection of potential groups $(G_1, \ldots, G_n)$ as output by potential recognition. The successful outcome is one group G from $(G_1, \ldots, G_n)$ together with a collective intention $C\text{-}INT_G(\varphi)$ among G to achieve φ, which includes corresponding individual intentions of all group members. This is achieved by subsequently attempting to establish the collective intention among G_1, G_2 etc. until this succeeds for some G_i. The collection of still untried potential groups $(G_{i+1}, \ldots, G_n)$ is stored for revision purposes.

Team formation is extensively treated in Chapter 8, with the focus on the relevant communication.

5.2.3   Plan Generation

Now that a strictly cooperative team has been created, it becomes time for the initiator to find a social plan realizing φ. The planning methods range from the use of a plan repository (Decker et al., 2001; Rao, 1996) to planning from first principles (de Silva et al., 2009; de Weerdt and Clement, 2009; desJardins et al., 1999; Durfee, 2008). Since this book is devoted to rather sophisticated forms of teamwork, we will describe plan generation for the latter, most complex kind of planning.

When planning from scratch in multi-agent settings, plan generation includes phases of *task division*, *means-end analysis* and *action allocation*. The input of this stage is a team G together with its collective intention $C\text{-}INT_G(\varphi)$. During successful planning, firstly an adequate task division of φ into a sequence of *subgoals* $\varphi_1, \ldots, \varphi_n$ is constructed, ensuring the certain realization of φ. Then, for each subgoal, means-end analysis provides *actions* realizing them. Finally, an appropriate *action allocation* to the team members is established and a temporal structure among the actions is devised. By 'appropriate allocation' we mean that not only agents' abilities and resources are considered, but also a proper composition of agent types in the team, as explained before. The result of this three-step process is a social plan P (see Section 4.3.1).

The strict planning phase is concluded by the distribution of commitments: all agents from the group socially commit to carry out their respective actions and communicate about these bilateral commitments to establish pair-wise mutual beliefs about them. In this way a proper level of awareness is reached in the team. All in all, a successful plan generation results in a collective commitment C-COMM$_{G,P}(\varphi)$ in the group G based on the social plan P. Paradigmatic forms of complex multi-agent interaction clearly come to the fore during plan generation, specifically deliberation, persuasion and negotiation. This takes place especially in cases where the stronger versions of collective commitment are considered.

5.2.4   Team Action

The previous three stages of teamwork finally culminate in team action, including both *execution of actions* and the *reconfiguration procedure*. In short, team members aim to realize their own projections (that is their individual actions) from the social plan. During this process many different situations may occur, some of which imply reconfiguration among the group. Reconfiguration was not treated explicitly in Wooldridge and Jennings (1996, 1999). We will discuss our approach to reconfiguration in the sequel.

Team action or plan realization succeeds when all actions making up the social plan P have been successfully executed by the committed agents and in this way the overall goal φ has been achieved. In the more realistic non-perfect case, some actions fail, requiring reconfiguration. As this may happen at any moment, the team action stage naturally includes the *reconfiguration process*. This process is carried out according to the *reconfiguration algorithm*, showing the phases of construction, maintenance and realization of collective commitments. Therefore, reconfiguration involves evolution of attitudes, formulated in terms of the four stages of teamwork and their complex interplay. The successful team action concludes the evolution of the team and its motivational attitudes.

5.3   The Reconfiguration Method

In the perfect case, a team achieves the goal in the way it was planned in the very beginning. When disturbances appear, however, some kind of *reconfiguration* is necessary. As indicated above, the main purpose of the reconfiguration algorithm is *the proper maintenance of collective commitments* on the team level and the associated social commitments and individual intentions on the individual level. As reconfiguration amounts to intelligent replanning, one needs to be aware of the current team's capabilities for planning: whether it is done from first principles or in a more standard way, like choosing plans from a plan repository.

To appropriately deal with the variety of situations or obstacles that appear during reconfiguration, the reasons of disturbances have to be recognized. The most common one is a failure of a certain action. When an action failed, a natural question is 'why'? However, answering this significant question might not be feasible, and, at any rate, does not solve the problem. Hence, it is much more conclusive to investigate whether anyone in the group is prepared to take over the failed action. If such an agent exists, we assume that the action has failed for a sort of *subjective reason*; therefore this problem may be relatively easily solved. As an example, an action may fail for a subjective reason if the committed agent no longer has resources for it, but a team colleague might still be prepared to pitch in and realize it.

Otherwise, when no one from the team is prepared to realize the action in the current state of the environment, there seem to be *objective reasons* for the action to fail. In computer science terminology this means that nobody can realize this action's pre-conditions.

Definition 5.1 *The execution of an action α fails for an* objective reason, *denoted as* $objective_G(\alpha)$, *if α is not realizable by anybody in the present team G in the current state of the environment that is,*
$$\neg \bigvee_{i \in G} can(i, G), \text{ which is short for } \neg \bigvee_{i \in G}(able(i, \alpha) \wedge opp(i, \alpha)).$$

Definition 5.2 *The execution of an action α fails for a* subjective reason, *denoted as* $subjective_G(\alpha)$ *if it fails, but not for an objective reason, that is,* $\bigvee_{i \in G} can(i, G)$.

Successful action realization leads to achieving its post-condition. In the reconfiguration algorithm below, during belief revision and motivational attitude revision in part **F**, care is taken to mark the agent that failed as unsuitable for α and not to try to reassign the failed action to this particular agent. If every failed action failed for a subjective reason, then action-reallocation based on the sequence of subjectively failed actions may be successful, and new social commitments to perform them may be created, calling for a revision of motivational attitudes.

The situation of objective reasons is much more complex, if not hopeless. The next step is to check how bad it is indeed, namely whether the action is *necessary* for achieving the overall goal, or might be appropriately substituted by another, hopefully achievable, one. The answer is related to the adopted planning method: when planning exploits a plan repository, it suffices to syntactically check whether the action in question appears in all relevant plans.

Definition 5.3 *Action α is* necessary *for a goal φ, if all applicable social plans P leading to post-condition φ (in the current state of the environment according to their pre-conditions), contain action α.*

On the other hand, when planning is done from first principles, the situation is much more complex. We somehow need to know whether the failed action is among the essential ones for a given goal. This is concerned with *semantic knowledge* related to the problem in question, which, however, can hardly be characterized a priori. If possible, it could be a good solution to have a list of actions necessary for a given goal, as a result from the planning phase. As we do not deal in detail with different types of planning, we will abstract from this problem-solving knowledge by introducing a predicate $necessary(\alpha, \varphi)$, stating whether the given action α is necessary for achieving the given goal φ.

Definition 5.4 *The failure of execution of action α blocks a goal φ, if α failed for an objective reason and α is necessary for φ, that is, $necessary(\alpha, \varphi)$ holds.*

This is the most serious negative case, generally leading to system-failure. Methods for checking whether failure of a certain action execution blocks the goal φ are again sensitive to the planning method and the application in question. Even though such checks might be viewed as an idealized notion, still they will be performed in the reconfiguration algorithm for efficiency reasons.

5.3.1   Continuity and Conservativity

When formulating the reconfiguration algorithm, we have chosen some intuitive properties corresponding to classical strategies adopted in backtracking. We postulate that the system behavior should preserve *continuity*. This means that, if an obstacle appears, the problems are solved by moving up in the hierarchy of teamwork stages, but as little as possible. The stages are ordered from potential recognition at the top to team action at the bottom. Thus, one moves to the nearest point up in the hierarchy of stages where a different choice is possible. If such a point does not exist anymore, the reconfiguration algorithm fails. In other words, depth-first search is used.

As regards generic stage-related procedures, a context-dependent question arises which 'local' results should be preferred. The answer calls for a domain-specific notion of the distance between teams, goals, plans, etc. At first glance, it seems that for a wide class of applications, it is justified that the system behaves in a *conservative* way (see also Nebel and Koehler, 1992). Importantly, conservativity (or *inertia*) entails that the collective commitment in question should change as little as necessary at every round of its evolution. The continuity criterion is application-independent and it determines the overall structure of the algorithm.

5.3.2   Reconfiguration Algorithm = Teamwork in Action

The reconfiguration algorithm presented below is meant to be generic: a pattern of behavior is described in terms of complex stage-associated procedures, called:

$$potential - recognition, team - formation, task - division,$$
$$means - end - analysis, action - allocation, plan - execution$$

without fixing any particular method or strategy. Input and output parameters, as well as other conditions of these procedures, are commented upon in the algorithm below (see Algorithm 1). As the environment is dynamic and often unpredictable, each of these procedures may succeed or fail – in this sense all stages of teamwork have a similar structure. Therefore, we use labels and appropriate GOTO statements for each stage-associated procedure to make the overall structure of the algorithm transparent. We introduce the predicate *succ* to denote that a procedure was performed successfully and the predicate *failed* to denote that a procedure was performed but failed. Note that in the reconfiguration algorithm, when the predicate *failed* occurs (say, in a line such as *failed*(division(φ))), then it does not cause any execution of the action that is its parameter (here division(φ)), it solely checks the status of the latest execution of this action; similarly for *succ*.

As indicated, the essence of the reconfiguration algorithm is the evolution of collective commitments and underlying plans. This inevitably leads to a *revision* of relevant individual, bilateral and collective motivational attitudes. In the algorithm, phases of belief revision and motivational attitude revision are distinguished, without further refinement, that is without splitting collective motivational attitudes into individual and social ones. They are realized by abstract procedures: `BeliefRevision` and `MotivationalAttitudeRevision`.

Pragmatically, the proper treatment of revision is ensured by the obligation that agents communicate about changes. On the other hand, the presence of social commitments and some form of awareness (for example common belief) about them *solely* between partners, together with the conservativity assumption, ensures that motivational attitudes revision is as efficient as possible.

The final outcome of the reconfiguration algorithm with respect to the overall goal φ is either system failure or success, realized by the generic procedures $system - failure(\varphi)$ and $system - success(\varphi)$, respectively. It is also assumed that all the required information is available at design-time.

In the algorithm (see Algorithm 1), the finite pool of agents T and H, denoting the collection of possible teams from T, are global parameters. Even though both refer to finite sets, it is possible that T and H evolve during the teamwork process; this makes reconfiguration suitable for an open environment.

Since the reconfiguration algorithm is formulated in a generic way, it needs to be tailored for specific applications. Undoubtedly, whatever domain is considered, the stage-associated procedures, including belief revision and motivational attitude revision, remain complex. The attitude revision during reconfiguration will be carefully treated in Chapter 6. Importantly, the algorithm structure is based on backtracking search, leaving room for improvements like informed search methods, including varieties of hill-climbing (Foss and Onder, 2006; Koza *et al.*, 2003) for particular applications.

5.3.3 Cycling through Reconfiguration

Reconfiguration is essentially a controlled type of evolution. To make this clear, let us focus on the failure points of the main stages of teamwork (see Algorithm 1):

1. The failure of potential recognition (see label **A**), meaning that agent a does not see any potential for cooperation with respect to the goal φ, leads to failure of the system.
2. The failure of team formation (see label **B**), meaning that the collective intention C-INT$_G(\varphi)$ cannot be established among any of the teams from H, requires a return to potential recognition to construct a new collection of potential teams.
3. The failure of task division (see label **C**) requires a return to team formation, in order to establish a collective intention in the chosen new team from H. This may be viewed as the reconfiguration of the team.
4. The failure of means-end analysis (see label **D**) requires a return to task division in order to create a new sequence of tasks, that would be the subject of a new round of means-end analysis.
5. The failure of action allocation (see label **E**) requires a return to means-end analysis in order to create a new sequence of actions that would be allocated to team members.

When, finally, a collective commitment is successfully established, the failure of some action executions from the social plan P leads to the evolution of the collective commitment, as a result of conservative replanning. This evolution will be discussed in Chapter 6.

Algorithm 1 Reconfiguration algorithm

A: $H \leftarrow$ `potential-recognition` (φ, a) ;
  if *failed* (`potential-recognition` (φ, a)) **then**
  | `system-failure` (φ) ;
  | STOP;
  $\{H = \{G_1, G_2,..., G_n\}$ - a collection of potential groups of agents is established$\}$
  $\{$PotCoop (φ, a, H) is established$\}$
B: $G \leftarrow$ `team-formation` (φ, a) ;
  if *failed* (`team-formation` (φ, a)) **then** GOTO A;
  $\{G$ - a team aiming to achieve $\varphi\}$
  $\{$C-INT$_G (\varphi)$ is established$\}$
C: $\sigma \leftarrow$ `division` (φ) ;
  if *failed* (`division` (φ)) **then**
  | $\{$`division` (φ) failed, return to `team-formation` (φ, a) in order to select
  | another group from H to create a team$\}$
  | GOTO B;
  $\{\sigma$ - a sequence of subgoals implying φ, that is the first part of a social plan $P\}$

D: $\tau \leftarrow$ `means-end-analysis` (σ) ;
  if *failed* (`means-end-analysis` (σ)) **then** GOTO C;
  $\{\tau$ - a sequence of actions resulting from σ, that is the second part of a social plan $P\}$

E: $P \leftarrow$ `action-allocation` (τ) ;
  if *failed* (`action-allocation` (τ)) **then** GOTO D;
  $\{P$ - a social plan for achieving $\varphi\}$

F: $(\tau_s, \tau_o) \leftarrow$ `plan-execution` (φ, G, P) ;
  $\{\tau_s$ - sequence of subjectively failed actions;
  τ_o - sequence of objectively failed actions$\}$

  if $\tau_s \cup \tau_o = \varnothing$ **then**
  | $\{$all actions from plan P are successfully executed; agents' beliefs, goals and intentions
  | need to be revised to reflect that φ is achieved$\}$
  | `BeliefRevision`;
  | `MotivationalAttitudeRevision`;
  | `system-success` (φ) ;
  | STOP;
  end
  else if $\tau_o = \varnothing$ **then**
  | $\{$`plan-execution` (φ, G, P) failed, there are no actions that failed for
  | objective reason and there are some that failed for subjective reason$\}$
  | **if** *succ* (`ActionReallocation` $(\varphi, G, \tau, \tau_s)$) **then**
  | | `MotivationalAttitudeRevision`;
  | | GOTO E;
  | **end**
  | **else** GOTO D;
  end
  else
  | $\{\tau_o \neq \varnothing\}$
  | $\{$`plan-execution` (φ, G, P) failed, there are some actions that failed for objective reason$\}$
  | `BeliefRevision`;
  | **if** `Blocked` (τ_o, φ) **then**
  | | `system-failure` (φ) ;
  | | STOP;
  | **end**
  | **else** GOTO D;
  end

5.3.4 Complexity of the Algorithm

The reconfiguration algorithm needs to be tailored to each application in question. This includes choosing adequate methods realizing stage-associated procedures. It is needless to say that especially plan generation is complex from both the AI and the MAS perspective. From the AI point of view, if planning is done from first principles, it is in general undecidable, while for limited domains it may still be tractable (Chapman, 1987). If, on the other hand, a plan library is used, searching the library may be complex as well (Nebel and Koehler, 1992, 1995). In fact, Nebel and Koehler (1995) show that in some circumstances, trying to find an existing plan for re-use is more complex than constructing a new one from scratch. From our daily life experience this conclusion is deeply true.

In advanced forms of planning, paradigmatic MAS interactions come to the fore (Durfee, 2008). The last years have highlighted that communication, negotiation (Kraus, 2001; Ramchurn et al., 2007, Rosenchein and Zlotkin, 1994) and coordination are indeed very complex. As regards dialogues involved in team formation, Chapter 8 will provide an informal glimpse into its complexity.

During team action, belief revision, which is known to be NP-hard, needs to be repeatedly performed. As explained in Chapter 2, recent times have seen a plethora of logical treatments of belief and knowledge revision. To this end, the complexity of belief revision is discussed in Benthem and Pacuit (2008). It is commonly understood that these advanced aspects have to be rigorously treated in *any* methodological approach. Still for specific application domains their complexity may be (much) lower than the worst cases above, for example whenever a restricted language may be used (see Chapter 9).

As a backtracking depth-first search is used during reconfiguration, iterative deepening may be applied to ascertain that the search method is complete and optimal. This does not significantly increase the time or space complexity when compared with depth-first search. Due to its exponential space complexity, however, breadth-first search is no option for reconfiguration.

As already indicated, our notion of collective commitment ensures efficiency of reconfiguration in two ways. Firstly, the motivational attitudes occurring in the definition are defined in a non-recursive way. This allows for their straightforward revision when necessary. Secondly, only bilateral commitments to actions are introduced, making the replanning and motivational attitudes revision even less complex. Wooldridge and Jennings (1996, 1999) define their *joint commitment* in a recursive way, based on the first step approaching the goal. In contrast, using our non-recursive concept of collective commitment, it often suffices to revise only the necessary pairwise commitments, limiting this way a scope of replanning.

5.4 Case Study of Teamwork: Theorem Proving

As reconfiguration is a highly complex issue, we will show an example of distributed problem solving in which the reconfiguration algorithm is implemented profitably. The overall goal of the system is to prove a new mathematical theorem. Even though the specification of the problem is not very detailed, for example, we do not give formal rules governing mathematical provability, we aim to highlight the main stages of teamwork.

5.4.1   Potential Recognition

We assume that all agents are prepared to work in a distributed environment, communicating, coordinating and negotiating when necessary. Suppose that the system starts with an open-minded initiator t who investigates the potential of cooperation to *prove*(*theorem*(T)) (see Section 4.3.4 for a definition of the open-minded commitment strategy). As a reminder, agents' goals and intentions can be formalized either with respect to a complex action such as *prove*(*theorem*(T)) or with respect to the corresponding state of affairs *proved*(*theorem*(T)); in the case study the action is chosen.

A general schema for proving the theorem has been formed by splitting it into a series of lemmas that together imply theorem T. Even though the initiator t has the goal that T be proved:

$$\text{GOAL}(t, prove(theorem(T)))$$

it cannot prove T on its own:

$$\neg can(t, prove(theorem(T)))$$

After communicating with agents l, p and c to find out about their willingness, abilities, opportunities and commitment strategies, agent t accepts the others as potential team members. Both l and p are single-minded agents of whom t believes that they can prove the lemmas needed for theorem T and c is a blindly committed agent of whom t believes that it can check all the proofs that will be constructed. These agents can be automated theorem provers or proof checkers or they can be viewed as human mathematicians supported by such programs. In the area of automated theorem proving, a multi-agent systems perspective has been shown to lead to effective and efficient solutions (Denzinger and Fuchs, 1999).

When recognizing the potential for cooperation, t considers the distribution of the agents' commitment characteristics. As discussed in Dunin-Kęplicz and Verbrugge (1996), blindly committed agents are not at all adaptive. Because scientific research necessitates frequent plan changes, it is crucial that the percentage of blindly committed agents in the team is not too high and that enough open-minded agents are present to pitch in if necessary.

Taking these requirements into account, $H = \langle\{t, l, c\}, \{t, p, c\}, \{t, l, p, c\}\rangle$ is the resulting collection of potential teams of which t believes that each team can jointly achieve the goal. In particular, for $G = \{t, l, c\}$, we have:

$$\text{BEL}(t, c\text{-}can_G(prove(theorem(T))))$$

All agents communicated to be willing, hence $\bigwedge_{i \in G}$ *willing* (i, *prove*(*theorem*(T))) holds. Moreover, all three potential teams contain only one blindly committed member and at least one open-minded agent, so they have an appropriate distribution of agent types:

$$\text{BEL}(t, propertypedistr(G, (prove(theorem(T)))))$$

In conclusion, PotCoop(*prove*(*theorem*(*T*)), *t*) is instantiated as follows:

PotCoop(*prove*(*theorem*(*T*)), *t*) ↔ GOAL(*t*, *prove*(*theorem*(*T*))) ∧

∃*G* ⊆ *H*   (BEL(*t*, *c-can*$_G$(*prove*(*theorem*(*T*)))) ∧ ⋀$_{i∈G}$ *willing* (*i*, *prove*(*theorem*(*T*))) ∧

propertypedistr(*G*, *prove*(*theorem*(*T*))))) ∧

(¬*can*(*t*, *prove*(*theorem*(*T*))) ∨ ¬GOAL(*t*, *done*(stit(*t*, *prove*(*theorem*(*T*)))))))

5.4.2   Team Formation

Now that the initiator has found potential team members, it needs to convince some of them to create a team. Therefore, *t* tries to establish the collective intention in the first team from *H*, for example *G* = {*t*, *l*, *c*}. Suppose that *t* succeeds and C-INT$_G$(*prove*(*theorem*(*T*))) is established. Otherwise, *t* successively attempts to establish the collective intention in the remaining potential teams.

5.4.3   Plan Generation

Assuming planning from first principles, plan generation consists of three stages: task division, means-end analysis and action allocation.

During *task division*, the team creates a sequence of subgoals σ = ⟨σ₁, σ₂⟩, with:

$$\sigma_1 = \text{``lemmas relevant for } T \text{ have been proved''} \text{ and}$$

$$\sigma_2 = \text{``theorem } T \text{ has been proved from lemmas''}$$

During *means-end analysis*, sequences of lemmas are determined from which the theorem *T* should be proved. Based on their mathematical knowledge, the agents construct lemma sequences that could in principle be true and that together lead adequately to the theorem. Suppose that there are different divisions $L_0, L_1, \ldots$ of the theorem into sequences of lemmas, where $L_i = (l_1^i, \ldots, l_{d(i)}^i)$. Suppose further that L_0 contains just two lemmas, lemma l_1^0 and lemma l_2^0. This determines the actions to be realized by the team:

$$\tau = \langle prove(lemma(l_1^0)), prove(lemma(l_2^0)), check(proof(l_1^0)), check(proof(l_2^0)),$$

$$prove(theorem\text{-}from(L_0)), check(proof\text{-}from(L_0))\rangle$$

See Table 5.1 for all actions necessary in the theorem-proving case study.

Table 5.1   Actions and their intended meanings.

prove(*theorem*(*T*))	To prove the theorem
prove(*lemma*(l_i^0))	To prove lemma l_i^0 from the sequence of lemmas $l_1^0, \ldots, l_n^0$
check(*proof*(l_i^0))	To check the proofs of the corresponding lemma l_i^0
prove(*theorem-from*(L_j))	To prove the theorem from the sequence of lemmas $L_j = \langle l_1^j, \ldots, l_{d(j)}^j \rangle$
check(*proof-from*(L_j))	To check all proofs of the theorem from the sequence of lemmas L_j

5.4.4   A Social Plan for Proving the Theorem

During *action allocation*, specific actions are allocated to agents having adequate abilities, opportunities, and commitment strategies. In this case, t (theorem prover) is responsible for proving the theorem T from the lemmas, l (lemma prover) is supposed to prove the needed lemmas, and c (proof checker) is given the task to check the others' proofs of lemmas and of the theorem. Hence, action allocation is clear and after creating a temporal structure, a social plan P is ready. In conclusion, the successful condition for proving T is as follows:

> There is a division of the theorem T into lemmas such that for each of them there exists a proof, constructed by the lemma prover and checked by the proof checker. Also, there is a proof of the theorem T from the lemmas, constructed by the theorem prover, which has been positively verified by the proof checker.

Assuming the sequential order of actions, this condition leads to the following social plan P to realize the team's goal *prove(theorem(T))*:

$$P = \langle \langle prove(lemma(l_1^0)), l \rangle; \langle check(proof\,(l_1^0)), c \rangle;$$
$$\langle prove(lemma(l_2^0)), l \rangle; \langle check(proof(l_2^0)), c \rangle;$$
$$\langle prove(theorem\text{-}from(L_0)), t \rangle; \langle check(proof\text{-}from(L_0)), c \rangle \rangle$$

In practice, a different temporal structure may be more efficient, including actions that are carried out in parallel, as in the examples of Sections 4.3.1 and 6.5.

5.4.5   A Collective Commitment to Prove the Theorem

When constructing the social plan, the team collectively makes sure that it is correct:

$$constitute\,(prove\,(theorem\,(T)), P)$$

and publicly establishes the following pairwise social commitments:

$$\text{COMM}(l, t, prove(lemma(l_1^0))) \wedge \text{COMM}(l, t, prove(lemma(l_2^0))) \wedge$$
$$\text{COMM}(t, l, prove(theorem\text{-}from(L_0))) \wedge \text{COMM}(c, l, check(proof(l_2^0))) \wedge$$
$$\text{COMM}(c, l, check(proof(l_2^0))) \wedge \text{COMM}(c, t, check(proof\text{-}from(L_0)))$$

Actually, no more social commitments are needed, although communication about the individual actions is always possible, for example in the context of scientific discussion. The common beliefs appearing in the definition of social commitments are established by bilateral communication between the agents involved.

Thanks to the collective planning, the team knows the plan and the corresponding pairwise commitments. Moreover, it commonly believes that the plan is correct. Since everything is under control, the team $G = \{t, l, c\}$ is prepared to establish a *robust collective commitment* (see Chapter 4):

$$\text{R-COMM}_{G,P}(prove(theorem(T))) \leftrightarrow \text{C-INT}_G(prove(theorem(T))) \wedge$$

$$constitute(prove(theorem(T)), P) \wedge \text{C-BEL}_G(constitute(prove(theorem(T)), P)) \wedge$$

$$\bigwedge_{i=1}^{2}(\text{COMM}(l, t, prove(lemma(l_i^0))) \wedge \text{COMM}(c, l, check(proof(l_i^0)))) \wedge$$

$$\text{COMM}(t, l, prove(theorem\text{-}from(L_0))) \wedge \text{COMM}(c, t, check(proof\text{-}from(L_0))) \wedge$$

$$\text{C-BEL}_G(\bigwedge_{i=1}^{2}(\text{COMM}(l, t, prove(lemma(l_i^0))) \wedge \text{COMM}(c, l, check(proof(l_i^0)))) \wedge$$

$$\text{COMM}(t, l, prove(theorem\text{-}from(L_0))) \wedge \text{COMM}(c, t, check(proof\text{-}from(L_0)))) \quad (5.1)$$

This robust collective commitment concludes both action allocation and plan generation.

5.4.6   Team Action

Below we sketch the reconfiguration procedure for an example. When an action fails, it is crucial to recognize why, because each reason gives rise to a different adequate reaction. For this case-study, we choose to make local decisions obeying conservativity, meaning that successful subproofs are not thrown away unnecessarily. Note, however, that in mathematical theorem proving in general, it is sometimes fruitful to take a completely new approach to a theorem, leading to a brand-new team plan.

During plan execution, the degree of responsibility and thus complexity of the three agents is quite different. The blindly committed agent works on its delegated actions without tracking what happens with the other agents. The single-minded and especially the open-minded agent have to react when the agents working for them fail in some actions. In short, when an agent fails to execute some actions, it responsibly communicates about this to its 'partner in commitment', who usually reacts. Suppose that in the current round, the team is proving the theorem T from lemma sequence $L_j = \langle l_1^j, \ldots, l_{d(j)}^j \rangle$. The reaction to failure in the reconfiguration algorithm is subdivided into two different cases, namely subjective and objective. First, cases **sub 1**, **sub 2** and **sub 3** are distinguished according to different *subjective reasons* for failure of actions:

sub 1 The proof checker c doesn't finish $check(proof(l_i^j))$ before a pre-set time limit, which naturally happens if c is an automated proof checker. In this case, the most efficient and conservative way of action reallocation is to let the single-minded lemma prover pitch in for c by checking its own proof and keep the rest of the plan unchanged. Thus, after belief revision, a new individual intention:

$$\text{INT}(l, check(proof(l_i^j)))$$

is added, as well as a social commitment:

$$\text{COMM}(l, c, check(proof(l_i^j)))$$

which is, finally, reflected in a new collective commitment. Note that c, being blindly committed, does not drop its social commitment to checking the proof.

sub 2 The lemma prover l does not believe he/she can *prove*$(lemma\,(l_i^j))$, but without finding a counterexample. The conservative way of action reallocation is to let t pitch in for l if t believes that it can prove the lemma itself:

$$\text{BEL}(t, can\,(t, prove\,(lemma\,(l_i^j))))$$

After belief revision, a single-minded agent l, drops its social commitment towards t and his/her individual intention to make the proof, which are both taken over by t:

$$\text{INT}(t, prove\,(lemma\,(l_i^j)))\ \text{and}\ \text{COMM}(t, l, prove\,(lemma\,(l_i^j)))$$

These are reflected in a new collective commitment.

sub 3 The theorem prover t does not believe that it can *prove*$(theorem\text{-}from\,(L_j))$, but without finding a counterexample. The conservative way of action reallocation is to let l pitch in for t if l believes that it can prove the theorem itself:

$$\text{BEL}(l, can\,(l, prove\,(theorem\text{-}from\,(L_j))))$$

After belief revision, agent t drops its social commitment towards l and its individual intention to make the proof, which are both taken over by l:

$$\text{INT}(l, prove\,(theorem\text{-}from\,(L_j)))\ \text{and}\ \text{COMM}(l, t, prove\,(theorem\text{-}from\,(L_j)))$$

These are reflected in a new collective commitment.

As regards *objective reasons* of failure, following the reconfiguration algorithm, it needs to be checked whether the overall goal *prove*$(theorem\,(T))$ is blocked. Here this happens solely when a counterexample to theorem T is found. Then the system fails.

There are also two non-blocking cases in which, following the reconfiguration algorithm, a new task division is needed:

ob 1 The proof checker c finds a mistake in a proof; then the proving has failed, but may be repeated by constructing a new proof of the same lemma or of the theorem. In this case, to obey conservativity, almost the same division of lemmas and the same task division, means-end analysis and action allocation are provided as in the present round. The only change is that the faulty prover gets as a new action to construct a new proof, which is then reflected in the social plan and in a social commitment: COMM($l, t, prove\text{-}lemma\text{-}diff\,(l_i^j, p)$) ($l$ commits to t to construct a new proof of lemma l_i^j, different from the faulty proof p) and finally in a new collective commitment.

ob 2 The proof checker c finds a counterexample to a lemma; then the action of proving has failed and the lemma is false, calling for a new round of task division. Suppose it was lemma l_i^j in the sequence of lemmas L_j that was found to be false. Then a new division of the theorem into a lemma sequence L_{j+1} has to be made. To obey conservativity, the new lemma sequence should resemble L_j as closely as possible. To this end, they especially try to conserve as many lemmas as possible from the initial sequence $l_1^j, \ldots, l_{i-1}^j$ which have already been proved and checked. Thus hard work is not thrown away unnecessarily and a new sequence of actions corresponding to L_{j+1} is created.

During action allocation, proving the lemmas is again allocated to l and proving the theorem from lemma sequence L_{j+1} is allocated to t. Agent c is again allocated to check all proofs. The resulting social plan is:

$$Q = \langle\langle prove(lemma(l_1^{j+1})), l\rangle; \langle check(proof\,(l_1^{j+1})), c\rangle; \ldots;$$

$$\langle prove(lemma(l_{d(j+1)}^{j+1})), l\rangle; \langle check(proof(l_{d(j+1)}^{j+1})), c\rangle;$$

$$\langle prove(theorem\text{-}from(L_{j+1})), t\rangle; \langle check(proof\text{-}from(L_{j+1})), c\rangle\rangle$$

Finally, a new robust collective commitment is made:

$$\text{R-COMM}_{G,Q}(prove(theorem(T))) \leftrightarrow \text{C-INT}_G(prove(theorem(T))) \wedge$$

$$\bigwedge_{i=1}^{d(j+1)} (\text{COMM}(l, t, prove(lemma(l_i^{j+1}))) \wedge$$

$$\text{COMM}(c, l, check(proof(l_i^{j+1})))) \wedge$$

$$\text{COMM}(t, l, prove(theorem\text{-}from(L_{j+1}))) \wedge$$

$$\text{COMM}(c, t, check(proof\text{-}from(L_{j+1}))) \wedge$$

$$\text{C-BEL}_G(\bigwedge_{i=1}^{d(j+1)} (\text{COMM}(l, t, prove(lemma(l_i^{j+1}))) \wedge$$

$$\text{COMM}(c, l, check(proof(l_i^{j+1})))) \wedge$$

$$\text{COMM}(t, l, prove(theorem\text{-}from(L_{j+1}))) \wedge$$

$$\text{COMM}(c, t, check(proof\text{-}from(L_{j+1}))))$$

Thus, both l and t revise their individual intentions and social commitments towards the lemmas that are not in the new sequence L_{j+1}, whereas c adds individual intentions and social commitments to l and t with respect to their new proofs.

After each reconfiguration step, the agents continue executing their actions until a new obstacle appears or the theorem is proved.

The theorem-proving example will be applied in a formal account of the evolution of commitments during reconfiguration in the next chapter. This formal aims to prove the correctness of the reconfiguration process.

6

The Evolution of Commitments during Reconfiguration

If you realize that all things change,
there is nothing you will try to hold on to.

Tao Te Ching (Lao-Tzu, Verse 74)

6.1 A Formal View on Commitment Change

The previous chapter covered a methodical approach to tackling the *reconfiguration problem*: when maintaining a collective intention during plan execution, it is crucial that agents re-plan properly and efficiently according to the situation changes. The essence of the reconfiguration algorithm is the dynamics of social and collective attitudes during teamwork. In a formal specification of these notions in BDI systems, different kinds of modal logics are exploited. Dynamic, temporal and epistemic logics are extensively used to describe the single agent case. Inevitably, social and collective aspects of teamwork should be investigated and formalized, again, in a combination of different kinds of modal logics.

As the static part of TEAMLOG, individual, social and collective attitudes have been defined in Chapters 2, 3 and 4. Now we aim to formally describe the maintenance of collective commitments during reconfiguration.

In case some action performance fails during team action, the realization of the collective commitment is threatened. Thus, some effects of the previous, potentially complex and resource consuming stages of teamwork may be wasted. However, to save the situation, often a minor correction of the social plan and the corresponding collective commitment suffices. For example, it might be enough to reallocate some actions to capable team members. If this is not feasible, a new plan may be established, slightly changing the existing one, etc. In the best case, the necessary changes are insignificant, re-using most of the previously obtained results.

As we argued in Chapter 5, the reconfiguration algorithm reflects a rigorous methodological approach to these changes, resulting in an *evolution* of collective commitment. In the current chapter, we characterize the properties of this process using dynamic logic.

Teamwork in Multi-Agent Systems: A Formal Approach   Barbara Dunin-Kęplicz and Rineke Verbrugge
© 2010 John Wiley & Sons, Ltd

This approach will allow us to precisely describe the results of complex actions and to highlight the change of motivational and informational attitudes during reconfiguration. Various aspects of evolution will be performed from a system developer's perspective, rather than the one of an agent. Finally, the properties describing system behavior, update or revision of agents' attitudes and properties of complex stage-related procedures provide a high-level specification of the system. This will enable a system designer to construct a correct system from the specification and to formally verify its behavior.

The reconfiguration procedure is based on the four-stage model of teamwork presented in Chapter 5. Before arguing formally that particular cases of reconfiguration are treated correctly, it has to be ensured that the stages are properly specified and then constructed. As a sort of an idealization, we introduce stage-associated procedures, viewed as *complex social actions*. These actions are highly context- and application-dependent. Therefore, they do not obey any generic axiomatization and a logical system characterizing them cannot be provided. Instead we formulate in an extended language of dynamic logic relevant high-level properties in the form of *semantic requirements*. These conditions should be ensured by the system developer when constructing a specific system (see Section 6.4).

6.1.1 Temporal versus Dynamic Logic

The effects of agents' individual actions and plans, as well as other changes in a system, can be modeled using variations of either dynamic logic or temporal logic. Because dynamic logic has been designed especially to represent reasoning about action and change, we decided to adopt the action-oriented dynamic logic approach here. Temporal logic is well-suited to describe more general changes over time, not those related to actions solely. Usually, BDI-logics are based on a linear or branching temporal logic, sometimes with some dynamic additions (Cohen and Levesque, 1990; Rao and Georgeff, 1991; Singh, 1990; Wooldridge, 2000). In the Appendix we present the temporal approach based on computation tree logic CTL (Emerson, 1990), that can be considered as an alternative to describe teamwork dynamics. This will give the interested reader a chance to compare the two approaches. Apparently, a full specification of the system includes complex temporal aspects, such as persistence of certain properties over time, usually until a given deadline. However, we will not introduce these procedural temporal elements into the logical framework. It is known that the combination of dynamic and temporal logic is extremely complex, especially in the presence of other modal operators, as is the case here (Benthem and Pacuit, 2006). Therefore, instead of making the logical system even more intractable and much harder to understand, we decided that temporal aspects should be left to the system developer to implement them in a procedural way.

This chapter is organized in the following way. In Section 6.2, the logical language and semantics are introduced. Section 6.3 is devoted to Kripke models and dynamic logic for actions and social plans. Section 6.4 gives a short overview of the four stages of teamwork. The central Section 6.5 presents in a multi-modal language how collective commitments evolve during reconfiguration. Finally, Section 6.6 focuses on discussion and options for further research.

6.2 Individual Actions and Social Plan Expressions

In this section we present the dynamic logic framework TEAMLOG^{dyn} reflecting dynamics of teamwork.

6.2.1 The Logical Language of TEAMLOG^{dyn}

Individual actions and formulas are defined inductively, both with respect to a fixed finite set of agents. The basis of the induction is given in the following definition.

Definition 6.1 (Language) The language is based on the following three sets:

- a denumerable set $\mathcal{P}$ of *propositional symbols*;
- a finite set $\mathcal{A}$ of *agents*, denoted by numerals $1, 2, \ldots, n$;
- a finite set $\mathcal{At}$ of *atomic actions*, denoted by a or b.

As indicated before, in TEAMLOG most modalities expressing agents' motivational attitudes appear in two forms: with respect to *propositions* reflecting a particular state of affairs, or with respect to *actions*. These actions are interpreted in a generic way – we abstract from any particular form of actions: they may be complex or primitive, viewed traditionally with certain effects or with default effects like in Dunin-Kęplicz and Radzikowska (1995a,b,c), etc. The transition from a proposition that agents intend to bring about to an action realizing this is achieved by means-end analysis, as discussed in Sections 5.2.3 and 6.4. The set of formulas (see Definition 6.5) is defined by a simultaneous induction, together with the set of individual actions $\mathcal{Ac}$, the set of complex social actions $\mathcal{Co}$ and the set of social plan expressions $\mathcal{Sp}$ (see Definitions 6.2, 6.3 and 6.4). The set $\mathcal{Ac}$ refers to agents' individual actions, usually represented without naming the agents, except when other agents are involved such as in **AC7** below. The individual actions may be combined into group actions by the social plan expressions defined below.

Below, we list operators to be used when defining individual actions and social plan expressions. However, the details of actions and social plans are not important for the purposes of this chapter, For example, another definition (for example without the iteration operation or without non-deterministic choice) may be used if more appropriate in a particular context.

Definition 6.2 (Individual actions) The set $\mathcal{Ac}$ of individual actions is defined inductively as follows:

AC1 each atomic action $a \in \mathcal{At}$ is an individual action;

AC2 if $\varphi \in \mathcal{L}$, then `confirm` (φ) is an individual action[1]; (Confirmation)

AC3 if $\alpha_1, \alpha_2 \in \mathcal{Ac}$, then $\alpha_1; \alpha_2$ is an individual action; (Sequential Composition)

AC4 if $\alpha_1, \alpha_2 \in \mathcal{Ac}$, then $\alpha_1 \cup \alpha_2$ is an individual action; (Non-Deterministic Choice)

AC5 if $\alpha \in \mathcal{Ac}$, then α^* is an individual action; (Iteration)

---

[1] In dynamic logic, `confirm`(φ) is usually denoted as $\varphi?$

AC6 if $\varphi \in \mathcal{L}$ and $i \in \mathcal{A}$, then $\texttt{stit}(\varphi)$ is an individual action;

AC7 if $\varphi \in \mathcal{L}$, $i, j \in \mathcal{A}$ and $G \subseteq \mathcal{A}$, then the following are individual actions: $\texttt{announce}_G(i, \varphi)$, $\texttt{communicate}(i, j, \varphi)$.

Recall that, in addition to the standard dynamic operators of **AC1** to **AC5**, the operator $\texttt{stit}$ of **AC6** is also introduced. The communicative actions $\texttt{announce}_G(i, \varphi)$ as well as $\texttt{communicate}(i, j, \varphi)$ and their role in updating beliefs of individuals and groups are discussed in Chapter 8.

For now, we assume the following. Given an agent i and an agent j, the action $\texttt{communicate}(i, j, \psi)$ stands for "agent i communicates to agent j that ψ holds". Next, given a group G and an agent $i \in G$, the action $\texttt{announce}_G(i, \psi)$ stands for "agent i announces to group G that ψ holds". Problems related to message delivery are disregarded in the sequel (but see Dunin-Kęplicz and Verbrugge (2005); Fagin *et al.*, (1995) and Van Baars and Verbrugge, (2007)).

Also recall that the complex social actions defined below refer to the four stages of teamwork, consecutively: ($\texttt{potential-recognition}$), ($\texttt{team-formation}$), ($\texttt{plan-generation}$) and ($\texttt{team-action}$).

Plan generation in turn is divided into three consecutive sub-stages, namely task division ($\texttt{division}$), means-end analysis ($\texttt{means-end-analysis}$) and action allocation ($\texttt{action-allocation}$). See Table 6.1 for all complex social actions and their intended meanings.

Table 6.1 Complex social actions and their intended meanings.

$\texttt{potential-recognition}(\varphi, H)$	Potential recognition in collection H of groups of agents towards goal φ
$\texttt{team-formation}(H, G)$	Formation of a team G from collection H of subsets of $\mathcal{A}$
$\texttt{plan-generation}(\varphi, G, P)$	Generation of a plan P leading to φ by the group G
$\texttt{division}(\varphi, \sigma)$	Division of the formula φ into a finite sequence of formulas σ
$\texttt{means-end-analysis}(\sigma, \tau)$	Means-end analysis assigning a finite sequence of individual actions τ to a finite sequence of formulas σ
$\texttt{action-allocation}(\tau, P)$	Allocation of individual actions from finite sequence τ in order to form a social plan expression P
$\texttt{system-success}(\varphi)$	Complex action performed in case of system success to achieve φ
$\texttt{system-failure}(\varphi)$	Complex action performed in case of system failure to achieve φ

Definition 6.3 (Complex social action) The set Co of complex social actions is introduced as follows:

C01 if φ is a formula, α is an individual action, $G \subseteq \mathcal{A}$, H a finite collection of subsets of $\mathcal{A}$, σ a finite sequence of formulas, τ a finite sequence of individual actions and P a social plan expression, then:

`potential-recognition`(φ, H), `team-formation`(H, G),

`plan-generation`(φ, G, P), `division`(φ, σ), `means-end-analysis`(σ, τ),

`action-allocation`(τ, P), `system-success`(φ), and `system-failure`(φ)

are complex social actions.

C02 If β_1 and β_2 are complex social actions, then so is $\beta_1; \beta_2$.

Definition 6.4 (Social plan expressions) The set $\mathcal{S}p$ of social plan expressions is defined inductively as follows:

SP1 If $\alpha \in \mathcal{A}c$ and $i \in \mathcal{A}$, then $\langle \alpha, i \rangle$ is a well-formed social plan expression;
SP2 If φ is a formula and $G \subseteq \mathcal{A}$, then $\text{stit}_G(\varphi)$ and $\text{confirm}(\varphi)$ are social plan expressions;
SP3 If α and β are social plan expressions, then $\langle \alpha; \beta \rangle$ (sequential composition) and $\langle \alpha \parallel \beta \rangle$ (parallellism) are social plan expressions.

The social plan $\text{confirm}(\varphi)$ (to test whether φ holds at the given world) is given here without group subscript, because the group does not influence the semantics (see Section 6.3.1). It will be clear from the context whether `confirm` is used as an individual action or as a social plan expression. Table 6.2 provides the relevant social plan expressions with their intended meanings.

Table 6.2   Social plan expressions and their intended meanings.

$\text{stit}_G(\varphi)$	The group G sees to it that φ
$\text{confirm}(\varphi)$	Plan to test whether φ holds at the given world

As to the modalities appearing in formulas below, see Section 6.3.1 for dynamic modalities, Chapter 2 for epistemic modalities and Chapters 3 and 4 for individual, social and collective motivational modalities.

Definition 6.5 (Formulas) We inductively define a set of formulas L as follows. (Table 6.3 below presents the relevant formulas with their intended meanings.)

F1 each atomic proposition $p \in \mathcal{P}$ is a formula;
F2 if φ and ψ are formulas, then so are $\neg\varphi$ and $\varphi \wedge \psi$;
F3 if φ is a formula, $\alpha \in \mathcal{A}c$ is an individual action, $\beta \in \mathcal{C}o$ is a complex social action, $i, j \in \mathcal{A}$, $G \subseteq \mathcal{A}$, σ a finite sequence of formulas, τ a finite sequence of individual actions and $P \in \mathcal{S}p$ a social plan expression, then the following are formulas:
epistemic modalities $\text{BEL}(i, \varphi)$, $\text{E-BEL}_G(\varphi)$, $\text{C-BEL}_G(\varphi)$;

Table 6.3   Formulas and their intended meanings.

$done\text{-}ac(i, \alpha)$	Action α has been performed by agent i
$succ\text{-}ac(i, \alpha)$	Action α has been successfully performed by agent i
$failed\text{-}ac(i, \alpha)$	Action α has been unsuccessfully performed by agent i
$done\text{-}sp(G, P)$	Social plan P has been performed by group G
$succ\text{-}sp(G, P)$	Social plan P has been successfully carried out by G
$failed\text{-}sp(G, P)$	Social plan P has been unsuccessfully carried out by G
$done\text{-}co(G, \beta)$	Complex social action β has been performed by G
$succ\text{-}co(G, \beta)$	Complex social action β was successfully performed by G
$failed\text{-}co(G, \beta)$	Complex social action β was unsuccessfully performed by G
$do\text{-}ac(i, \alpha)$	Agent i is just about to perform action α
$do\text{-}sp(G, P)$	Group G is about to start carrying out plan P
$do\text{-}co(G, \beta)$	Group G is about to start performing complex social action β
$able(i, \alpha)$	Agent i is able to realize action α
$opp(i, \alpha)$	Agent i has the opportunity to realize action α
$[do(i, \alpha)]\varphi$	After performing action α by agent i, φ holds
$[\beta]\varphi$	After performing complex social action β, φ holds
$[P]\varphi$	After carrying out plan P, φ holds
$division(\varphi, \sigma)$	σ is the sequence of subgoals resulting from decomposition of φ
$means(\sigma, \tau)$	τ is the sequence of actions resulting from means-end analysis on σ
$allocation(\tau, P)$	P is a social plan resulting from allocating the actions from τ to interested team members
$constitute(\varphi, P)$	P is a correctly constructed social plan to achieve φ

motivational modalities $GOAL(i, \varphi)$, $GOAL(i, \alpha)$, $INT(i, \varphi)$, $INT(i, \alpha)$, $COMM(i, j, \varphi)$, $COMM(i, j, \alpha)$, $E\text{-}INT_G(\varphi)$, $E\text{-}INT_G(\alpha)$, $M\text{-}INT_G(\varphi)$, $M\text{-}INT_G(\alpha)$, $C\text{-}INT_G(\varphi)$, $C\text{-}INT_G(\alpha)$, $S\text{-}COMM_{G,P}(\varphi)$, $S\text{-}COMM_{G,P}(\alpha)$;

execution modalities $done\text{-}ac(i, \alpha)$, $succ\text{-}ac(i, \alpha)$, $failed\text{-}ac(i, \alpha)$; $done\text{-}sp(G, P)$, $succ\text{-}sp(G, P)$, $failed\text{-}sp(G, P)$; $done\text{-}co(G, \beta)$, $succ\text{-}co(G, \beta)$, $failed\text{-}co(G, \beta)$, $do\text{-}ac(i, \alpha)$, $do\text{-}sp(G, P)$, $do\text{-}co(G, \beta)$;

abilities and opportunities $able(i, \alpha)$, $opp(i, \alpha)$;

dynamic modalities $[do(i, \alpha)]\varphi$, $[\beta]\varphi$, $[P]\varphi$;

stage results $division(\varphi, \sigma)$, $means(\sigma, \tau)$, $allocation(\tau, P)$, $constitute(\varphi, P)$.

The stage results in the above definition refer to the results of the three substages of plan generation, namely task division, means-end analysis and action allocation. The predicate $constitute(\varphi, P)$ (for "P constitutes a correctly constructed social plan for realizing state of affairs φ") is defined in Section 6.4.1.

The constructs $\bot$, $\vee$, $\rightarrow$ and $\leftrightarrow$ are defined in the usual way.

6.3   Kripke Models

Let us introduce the most extensive Kripke models of this book, extending those of Section 4.2 with a dynamic component. Each Kripke model for the language defined in the previous section consists of a set of worlds, a set of accessibility relations between worlds and a valuation of the propositional atoms, as follows. The definition also includes semantics for derived operators corresponding to abilities, opportunities and performance of (individual or social) actions.

Definition 6.6 (Kripke model) A Kripke model is a tuple:

$$\mathcal{M} = (W, \{B_i : i \in \mathcal{A}\}, \{G_i : i \in \mathcal{A}\}, \{I_i : i \in \mathcal{A}\},$$

$$\{R_{i,\alpha} : i \in \mathcal{A}, \alpha \in Ac\}, \{R_\beta : \beta \in Co\}, \{R_P : P \in Sp\};$$

$$Val, abl, op, perfac, perfsp, perfco, nextac, nextco, nextsp);$$

such that:

1. W is a set of possible worlds, or states.
2. For all $i \in \mathcal{A}$, it holds that $B_i, G_i, I_i \subseteq W \times W$. They stand for the accessibility relations for each agent w.r.t. beliefs, goals and intentions, respectively.
3. For all $i \in \mathcal{A}$, $\alpha \in Ac$, $\beta \in Co$ and $P \in Sp$, it holds that $R_{i,\alpha}, R_\beta, R_P \subseteq W \times W$. They stand for the dynamic accessibility relations.[2]
4. $Val : \mathcal{P} \times W \to \{0, 1\}$ is the function that assigns the truth values to propositional formulas in states.
5. $abl : \mathcal{A} \times Ac \to \{0, 1\}$ is the ability function such that $abl(i, \alpha) = 1$ indicates that agent i is able to realize the action α. $\mathcal{M}, v \models able(i, \alpha) \Leftrightarrow abl(i, \alpha) = 1$.
6. $op : \mathcal{A} \times Ac \to (W \to \{0, 1\})$ is the opportunity function such that $op(i, \alpha)(w) = 1$ indicates that agent i has the opportunity to realize action α in world w. $\mathcal{M}, v \models opp(i, \alpha) \Leftrightarrow op(i, \alpha)(v) = 1$.
7. $perfac : \mathcal{A} \times Ac \to (W \to \{0, 1, 2\})$ is the individual action performance function such that $perfac(i, \alpha)(w)$ indicates the result in world w of the performance of individual action α by agent i in world w (here, 0 stands for failure, 1 for success and 2 stands for "undefined", for example if w is not the endpoint of an $R_{(i,\alpha)}$ accessibility relation).
   - $\mathcal{M}, v \models succ\text{-}ac(i, \alpha) \Leftrightarrow perfac(i, \alpha)(v) = 1$.
   - $\mathcal{M}, v \models failed\text{-}ac(i, \alpha) \Leftrightarrow perfac(i, \alpha)(v) = 0$.
   - $\mathcal{M}, v \models done\text{-}ac(i, \alpha) \Leftrightarrow perfac(i, \alpha)(v) \in \{0, 1\}$.
8. $perfco : 2^{\mathcal{A}} \times Co \to (W \to \{0, 1, 2\})$ is the complex social action performance function such that $perfco(j, \beta)(w)$ indicates the result in world w of the performance of complex social action β by a group of agents j.
   - $\mathcal{M}, v \models succ\text{-}co(j, \beta) \Leftrightarrow perfco(j, \beta)(v) = 1$.
   - $\mathcal{M}, v \models failed\text{-}co(j, \beta) \Leftrightarrow perfco(j, \beta)(v) = 0$.
   - $\mathcal{M}, v \models done\text{-}co(j, \beta) \Leftrightarrow perfco(j, \beta)(v) \in \{0, 1\}$.
9. $perfsp : 2^{\mathcal{A}} \times Sp \to (W \to \{0, 1, 2\})$ is the social plan performance function such that $perfsp(j, P)(w)$ indicates the result in world w of the performance of social plan P by a group of agents j.
   - $\mathcal{M}, v \models succ\text{-}sp(j, P) \Leftrightarrow perfasp(j, P)(v) = 1$.
   - $\mathcal{M}, v \models failed\text{-}sp(j, P) \Leftrightarrow perfsp(j, P)(v) = 0$.
   - $\mathcal{M}, v \models done\text{-}sp(j, P) \Leftrightarrow perfsp(j, P)(v) \in \{0, 1\}$.

---

[2] For example, $(w_1, w_2) \in R_{i,\alpha}$ means that w_2 is a possible resulting state from w_1 by i executing action α.

10. *nextac* : $\mathcal{A} \times \mathcal{A}c \to (W \to \{0, 1\})$ is the next moment individual action function such that *nextac*$(i, \alpha)(w)$ indicates that in world w agent i will next perform action α. $\mathcal{M}, v \models \textit{do-ac}(i, \alpha) \Leftrightarrow \textit{nextac}(i, \alpha)(v) = 1$.

11. *nextco* : $2^{\mathcal{A}} \times \mathcal{C}o \to (W \to \{0, 1\})$ is the next moment complex social action performance function such that *nextco*$(j, \beta)(w)$ indicates that in world w the group of agents j will next start performing the complex social action β. $\mathcal{M}, v \models \textit{do-co}(j, \beta) \Leftrightarrow \textit{nextco}(j, \beta)(v) = 1$.

12. *nextsp* : $2^{\mathcal{A}} \times \mathcal{S}p \to (W \to \{0, 1\})$ is the next moment social plan performance function such that *nextsp*$(j, P)(w)$ indicates that in world w the group of agents j will next start performing social plan P. $\mathcal{M}, v \models \textit{do-sp}(j, P) \Leftrightarrow \textit{nextsp}(j, P)(v) = 1$.

Both abilities and opportunities are modeled in the above definition in a static way. Apparently, more refined definitions, using a language that includes dynamic and/or temporal operators (see for example Brown (1988); Dunin-Kęplicz and Radzikowska (1995b); van Linder *et al.* (1998)) are possible. We have chosen not to do so here, because these concepts are not the main focus of this chapter. We do assume that the functions are in accord with the construction of complex individual actions, for example, if an agent is able to realize $a; b$, then it is able to realize a. Similarly, we have modeled action performance for individual actions, social plans and complex social actions by functions (*perfac*, *perfsp* and *perfco*), addressing the question whether a certain action has just been performed, and if so, whether it was successful. Finally the functions *nextac*, *nextsp* and *nextco* model whether a certain action will be executed next. Again, these functions are assumed to agree with the construction of complex actions, for example, if *perfac*$(i, a; b) = 1$, then *perfac*$(i, b) = 1$.

We use three-valued performance functions for actions, complex social actions and social plan expressions, because at many worlds it may be that the relevant action has not been performed at all. Of course one could also use partial functions here (where our value 2 is replaced by "undefined").

The truth conditions pertaining to the propositional part of the language $\mathcal{L}$ are the standard ones used in modal logics. The derived operators above correspond in a natural way to the results of the ability, opportunity, performance and next execution functions. For example, $\mathcal{M}, v \models \textit{done-sp}(j, P)$ is meant to be true if team j just executed the social plan P, as modelled by the performance function giving a value other than 2 (undefined), that is *perfsp*$(j, P)(v) \in \{0, 1\}$.

In the remainder of the chapter we will mostly abbreviate all the above forms of success, failure and execution (past and future) for actions, complex actions and social plans to simply *succ*, *failed*, *done* and *do*.

The truth conditions for formulas with dynamic operators as main modality are given in Section 6.3.1, for those with epistemic main operators, the truth definitions are given in Chapter 2 and finally finally, for those with motivational modalities as main operators, the definitions are given in Chapter 3.

6.3.1 Axioms for Actions and Social Plans

In the semantics, the relations $R_{i,a}$ for atomic actions a are given. The other accessibility relations $R_{i,\alpha}$ for actions are built up from these as follows in the usual way (Harel *et al.*, 2000).

Definition 6.7 (Dynamic accessibility relations for actions)

- $(v, w) \in R_{i,\texttt{confirm}(\varphi)} \Leftrightarrow (v = w$ and $\mathcal{M}, v \models \varphi)$;
- $(v, w) \in R_{i,\alpha_1;\alpha_2} \Leftrightarrow \exists u \in W[(v, u) \in R_{i,\alpha_1}$ and $(u, w) \in R_{i,\alpha_2}]$;
- $(v, w) \in R_{i,\alpha_1 \cup \alpha_2} \Leftrightarrow [(v, w) \in R_{i,\alpha_1}$ or $(v, w) \in R_{i,\alpha_2}]$;
- R_{i,α^*} is the reflexive and transitive closure of $R_{i,\alpha}$.

In a similar way, the accessibility relations for social plan expressions and complex social actions are built up from those of individual actions in an appropriate way (Harel *et al.*, 2000; Peleg, 1987). We do not give the complete definition, but for example, we have:

- If $\alpha \in \mathcal{Ac}$ and $i \in \mathcal{A}$, then $(v, w) \in R_{\langle \alpha, i \rangle} \Leftrightarrow (v, w) \in R_{i,\alpha}$.

Now we can define the valuations of complex formulas containing dynamic operators as the main operator.

Definition 6.8 (Valuation for dynamic operators)   Let φ be a formula, $i \in \mathcal{A}$, $\alpha \in \mathcal{Ac}$, $\beta \in \mathcal{Co}$, and $P \in \mathcal{Sp}$.

actions $\mathcal{M}, v \models [do(i, \alpha)]\varphi \Leftrightarrow$ for all w with $(v, w) \in R_{i,\alpha}, \mathcal{M}, w \models \varphi$.
social plan expressions $\mathcal{M}, v \models [P]\varphi \Leftrightarrow$ for all w with $(v, w) \in R_P, \mathcal{M}, w \models \varphi$.
complex social actions $\mathcal{M}, v \models [\beta]\varphi \Leftrightarrow$ for all w with $(v, w) \in R_\beta, \mathcal{M}, w \models \varphi$.

For the dynamic logic of actions, we adapt the axiomatization PDL of propositional dynamic logic, as found in Goldblatt (1992):

P2 $[do(i, \alpha)](\varphi \rightarrow \psi) \rightarrow ([do(i, \alpha)]\varphi \rightarrow [do(i, \alpha)]\psi)$;   (Dynamic Distribution)
P3 $[do(i, \texttt{confirm}(\varphi))]\psi \leftrightarrow (\varphi \rightarrow \psi)$;
P4 $[do(i, \alpha_1; \alpha_2)]\varphi \leftrightarrow [do(i, \alpha_1)][do(i, \alpha_2)]\varphi$;
P5 $[do(i, \alpha_1 \cup \alpha_2)]\varphi \leftrightarrow ([do(i, \alpha_1)]\varphi \wedge [do(i, \alpha_2)]\varphi$;
P6 $[do(i, \alpha^*)]\varphi \rightarrow \varphi \wedge [do(i, \alpha)][do(i, \alpha^*)]\varphi$;   (Mix)
P7 $(\varphi \wedge [do(i, \alpha^*)](\varphi \rightarrow [do(i, \alpha)]\varphi)) \rightarrow [do(i, \alpha^*)](\varphi)$;   (Induction)
PR2 From φ, derive $[do(i, \alpha)]\varphi$.   (Dynamic Necessitation)

The axiom system PDL is sound and complete with respect to Kripke models with only the dynamic accessibility relations $R_{i,\alpha}$ as defined above. Its decision problem is exponential time complete, as proved by Fischer and Ladner (1979).

One needs to add axioms for complex social actions and social plan expressions in an appropriate way, for example, for all $\mathcal{M}, w$ and the group version of $\texttt{confirm}$:

$$\mathcal{M}, w \models [\texttt{confirm}(\psi)]\chi \leftrightarrow (\psi \rightarrow \chi)$$

As this is not the main subject of this chapter and as the axiom systems depend on the domain in question, we will not include a full system here. However, for the ‖-operator, one may use the appropriate axioms for concurrent dynamic logic as found in Harel *et al.* (2000).

For the notation on informational and motivational attitudes relevant to this chapter, we refer the reader to Tables 2.1, 3.1, 3.2 and 4.1 in Chapters 2, 3 and 4, respectively.

6.4 Dynamic Description of Teamwork

In Chapter 4, we presented a tuning machine allowing us to define different versions of collective commitments, reflecting different aspects of teamwork and applicable in different situations. Strong collective commitment has been calibrated to express the flavor and strength of a group commitment that is applicable in many situations. Therefore we have chosen it as the exemplary type to illustrate the evolution of collective commitment:

$$S\text{-COMM}_{G,P}(\varphi) \leftrightarrow C\text{-INT}_G(\varphi) \wedge$$

$$constitute(\varphi, P) \wedge C\text{-BEL}_G(constitute(\varphi, P)) \wedge$$

$$\bigwedge_{\alpha \in P} \bigvee_{i,j \in G} COMM(i, j, \alpha) \wedge C\text{-BEL}_G(\bigwedge_{\alpha \in P} \bigvee_{i,j \in G} COMM(i, j, \alpha))$$

See Section 4.5.2 for more explanations and an example of strong collective commitment. Also, recall that teams of agents have positive introspection about strong collective commitments among them, even if negative introspection does not follow from the defining axiom (see Section 4.5.6).

6.4.1 Operationalizing the Stages of Teamwork

Now we go on to specify a formal system realizing the four stages of teamwork: *potential recognition, team formation, plan formation* and *team action*. To this end we assume that these generic stages are realized by black box-like abstract procedures: `potential-recognition`, `team-formation`, `division`, `means-end-analysis` and `action-allocation`. Formally, they are viewed as complex social actions that will not be further refined. This job belongs to the system developer building a system for a given application.

When maintaining a collective commitment and its constituent collective intention, social commitments and individual intentions during plan execution, it is crucial that agents re-plan properly and efficiently when some actions fail during execution. This is the essence of the *reconfiguration algorithm* presented in Chapter 5. The current chapter carefully treats the evolution of collective commitments during reconfiguration. To this end, we build a formal system based on a generic reconfiguration algorithm. Thus, for all four stages of teamwork, both the positive case (when the stage-associated action succeeds) and the negative case (when this action fails) will be specified, and treated accordingly. In this way, the appropriate properties of the system will be formulated during design time. Their realization should be ensured by the system developer during run time.

6.4.1.1 Potential Recognition

Analogous to Wooldridge and Jennings (1999), we consider teamwork to begin when some agent recognizes the potential for cooperative action in order to reach its goal. As a reminder, the input of this stage is an initiator agent a, a goal φ plus a finite set $T \subseteq \mathcal{A}$ of agents from which a potential team may be formed. The output is the "potential for cooperation" $PotCoop(\varphi, a)$ (see Section 5.2.1) meaning that agent a sees a potential to realize φ and a collection $H = (G_1, \ldots, G_n)$ of potential teams is constructed.

Potential recognition is realized by a complex action `potential-recognition`. In the case of its successful performance by agent a we have:

Ps

$$succ(\texttt{potential-recognition}(\varphi, a)) \rightarrow \text{PotCoop}(\varphi, a)$$

Otherwise, the failure of `potential-recognition`, meaning that agent a doesn't see any potential to achieve φ, leads to the failure of the system:

Pf

$$failed(\texttt{potential-recognition}(\varphi, a)) \rightarrow do(\texttt{system-failure}(\varphi))$$

In the notation for results of actions inspired by dynamic logic, this stands for: after potential recognition has failed, then the whole system fails to achieve φ and the action `system-failure`(φ) is to be done. The `system-failure`(φ) and `system-success`(φ) are realized by complex actions, to be refined by the system developer. At any rate, this implies that any further activity towards achieving φ is stopped. See Section 6.3 for a formal explanation of the concepts *do* and *failed*.

6.4.1.2 Team Formation

Suppose that agent a sees the potential for cooperation to achieve φ. During team formation a attempts to establish in some team G the *collective intention* C-INT$_G(\varphi)$. The input of this stage is agent a, goal φ and a collection $H = (G_1, \ldots, G_n)$ resulting from potential recognition. The successful outcome is *one* team G from H together with a collective intention C-INT$_G(\varphi)$. Assume that team formation is realized by a complex action `team-formation`. After its successful performance we have:

Ts

$$succ(\texttt{team-formation}(\varphi, a, G)) \rightarrow \text{C-INT}_G(\varphi)$$

Otherwise, the failure of `team-formation`, meaning that the collective intention cannot be established among any of the teams from H, requires the return to potential recognition to obtain a new collection of potential teams:

Tf

$$failed(\texttt{team-formation}(\varphi, a, G)) \rightarrow do(\texttt{potential-recognition}(\varphi, a))$$

The reasons for failure of team formation are usually situation-dependent, typical ones dealing with unwillingness or overcommitment of agents.

6.4.1.3 Plan Generation

During team formation, a loosely coupled group of agents has become a strictly cooperative team. Now the team action needs to be prepared in detail. This entails planning, resulting in a social plan P and, roughly speaking, agents accepting commitments towards actions from P, leading to a collective commitment. At that point the team is ready to call "Action now!"

Before reaching this moment, the input of plan generation is a team G with collective intention C-INT$_G(\varphi)$ towards φ. The successful outcome is a strong collective commitment S-COMM$_{G,P}(\varphi)$ based on the newly generated social plan P.

6.4.1.4 Properties of Planning

We see planning in TEAMLOG as a three-step process. The first step is *task division*, in which a decomposition of a complex task φ into (possibly also complex) subgoals is addressed. This phase is performed by a complex action `division` resulting in a predicate *division*(φ, σ) standing for "the sequence σ is a result of task decomposition of goal φ into subgoals". Here, σ is a finite sequence of propositions standing for goals, for example $\sigma = \langle \varphi_1, \ldots, \varphi_n \rangle$. Thus, successful performance of task division leads to *division*(φ, σ). This is formalized using dynamic logic as follows:

Ds

$$succ(\texttt{division}(\varphi, \sigma)) \rightarrow division(\varphi, \sigma)$$

Otherwise, we have:

Df

$$failed(\texttt{division}(\varphi, \sigma)) \rightarrow \neg division(\varphi, \sigma)$$

Next follows *means-end analysis*, aiming to determine appropriate means leading to the ends, that is actions realizing particular subgoals. Note that for any subgoal there may be many (possibly complex) actions to achieve it. Again, we assume that this phase is realized by a complex action `means-end-analysis`, resulting in a predicate *means*(σ, τ), standing for "the action sequence τ is a result of means-end analysis for the subgoal sequence σ". Here, τ is a finite sequence of actions $\in \mathcal{A}c$, for example $\tau = \langle \alpha_1, \ldots, \alpha_n \rangle$. This is a generalization of the standard one-step process, which is performed for a single goal at a time. Note that to each subgoal in σ a number of actions may be associated, so that σ and τ may have different lengths. Thus, the successful result of means-end analysis is *means*(σ, τ):

Ms

$$succ(\texttt{means-end-analysis}(\sigma, \tau)) \rightarrow means(\sigma, \tau)$$

Otherwise, we have:

Mf

$$failed(\texttt{means-end-analysis}(\sigma, \tau)) \rightarrow \neg means(\sigma, \tau)$$

Next follows *action allocation*, in which the actions resulting from means-end analysis are given to interested team members. It is realized by a complex action `action-allocation`. This results first in pairs $\langle \alpha, i \rangle$ of an action α and an agent i. To make allocation complete, the temporal structure among pairs $\langle \alpha, i \rangle$ needs to be established. The construction of a social plan is formalized by the predicate $allocation(\tau, P)$, standing for "P is a social plan resulting from allocation a sequence of actions τ to interested team members". Thus, the successful result of action allocation is represented formally by:

As

$$succ(\texttt{action-allocation}(\tau, P)) \rightarrow allocation(\tau, P)$$

Otherwise, we have:

Af

$$failed(\texttt{action-allocation}(\tau, P)) \rightarrow \neg allocation(\tau, P)$$

Note that in the predicates *division*, *means* and *allocation*, it does not matter that the lengths of the subgoal sequence and the action sequence are not fixed in advance; one can always code finite sequences in such a way that their length may be recovered from the code.

6.4.1.5 Properties of the Predicate $constitute(\varphi, P)$

After formally describing the phases of plan generation, it is time to give a formal meaning to the predicate *constitute* that made its first informal appearance in Chapter 4. As a reminder, the predicate $constitute(\varphi, P)$ informally stands for "P is a correctly constructed social plan to achieve φ". A collective commitment to achieve φ based on plan P always contains the component $constitute(\varphi, P)$, formally defined as:

C0

$$constitute(\varphi, P) \leftrightarrow$$
$$\exists \sigma \exists \tau (division(\varphi, \sigma) \wedge means(\sigma, \tau) \wedge allocation(\tau, P))$$

All in all, the overall planning consists of the complex action `division; means-end-analysis; action-allocation`. During its successful performance a correct plan is constructed:

C1

$$succ(\texttt{division}(\varphi, \sigma); \texttt{means-end-analysis}(\sigma, \tau);$$
$$\texttt{action-allocation}(\tau, P)) \rightarrow constitute(\varphi, P)$$

Ultimately, the successful realization of the plan P ensures the achievement of φ:

CS

$$constitute(\varphi, P) \rightarrow [\texttt{confirm}(succ(P))]\varphi$$

Moreover, if planning is performed collectively, especially from first principles, the plan is known to all team members, as reflected in the conjunct C-BEL$_G$($constitute(\varphi, P)$).

The failure of planning will now be considered more in detail, looking carefully at which step the failure actually takes place.

Thus, the failure of task-division, meaning that no task division for φ was found, requires a return to team formation to choose a new team. It may be viewed as reconfiguration of the team together with revision of the collective intention and the respective individual attitudes.

Dd

$$failed(\texttt{division}(\varphi, \sigma)) \rightarrow do(\texttt{team-formation}(\varphi, a, G'))$$

where $G' \in H$ denotes a new potential team.

The failure of means-end-analysis, meaning that there are no available means to realize some subgoals from a goal sequence σ, requires a return to the task division stage to construct a new sequence of subgoals.

Md

$$failed(\texttt{means-end-analysis}(\sigma, \tau)) \rightarrow do(\texttt{division}(\varphi, \sigma'))$$

where σ' denotes a new sequence of subgoals devised to realize φ.

The failure of action-allocation, meaning that some of the previously established actions cannot be allocated to agents in G, requires a return to means-end analysis for new means (that is actions) that could be allocated to members of the current team.

Ad

$$failed(\texttt{action-allocation}(\tau, P)) \rightarrow do(\texttt{means-end-analysis}(\sigma, \tau'))$$

where τ' denotes a new sequence of actions selected to achieve the subgoals in σ.

In the last two cases, when backtracking is considered, some previously obtained partial results may be re-used to establish $constitute(\varphi, P')$ for a new social plan P'. In this way *conservativity* is maintained (see Chapter 5 for explanations). Thus, in addition to the obvious **C1**, the following holds:

C2

$$division(\varphi, \sigma) \wedge means(\sigma, \tau) \wedge succ(\texttt{action-allocation}(\tau, P')$$
$$\rightarrow constitute(\varphi, P')$$

C3

$$division(\varphi, \sigma) \wedge succ(\texttt{means-end-analysis}(\sigma, \tau');$$

$$\texttt{action-allocation}(\tau', P)) \rightarrow constitute(\varphi, P')$$

In fact, **C2** follows directly from **C0** and **As**.

6.4.1.6 Construction of $S\text{-}COMM_{G,P}(\varphi)$ by Communication

So far a correct plan leading to φ has been developed. It is now time to establish a relevant collective commitment based on it. Actually, procedures applicable to different types of group commitment vary significantly, including a variety of communicative acts and/or communication protocols. Instead of exploring them in detail, we focus on strong commitment $S\text{-}COMM_{G,P}$, reflecting typical aspects of cooperation in real-life situations. To establish it, three phases of communication will be formally characterized in terms of their results.

The communication starts when:

(i) $C\text{-}INT_G(\varphi)$ and
(ii) $constitute(\varphi, P)$

are in place as a result of previous stages. Since we deal with $S\text{-}COMM_{G,P}(\varphi)$;

(iii) $C\text{-}BEL_G(constitute(\varphi, P))$

has to be established. After the social plan is communicated, agents from a team need to *socially commit* to carry out respective actions, and to communicate these decisions in order to establish pairwise mutual beliefs about them. This phase will be concluded with both $COMM(i, j, \alpha)$ for all actions α from P as well as a collective belief that relevant commitments have been made. Again, the above phases may be handled by different communication protocols. By means of the chosen one, first

(iv) $\bigwedge_{\alpha \in P} \bigvee_{i,j \in G} COMM(i, j, \alpha)$

needs to be in place, after which a consecutive phase of communication should lead to

(v) $C\text{-}BEL_G(\bigwedge_{\alpha \in P} \bigvee_{i,j \in G} COMM(i, j, \alpha))$.

Finally, after the appropriate exchange of information together with (i), (ii) and (iii), a strong collective commitment $S\text{-}COMM_{G,P}(\varphi)$ in a team G, based on plan P towards φ holds.

Let `construction` be the application-dependent complex social action establishing (iv) and (v), obeying the following postulate:

CTR

$$C\text{-}INT_G(\varphi) \wedge constitute(\varphi, P) \wedge succ(\texttt{construction}(\varphi, G, P))$$

$$\rightarrow S\text{-}COMM_{G,P}(\varphi)$$

This information exchange concludes the *collective* part of planning and the team is ready to start *team action*. Section 6.5 treats what happens in a dynamic environment, where some actions from the social plan fail.

6.4.1.7 The Frame Problem

The *frame problem* in artificial intelligence was formulated by McCarthy and Hayes (1969). It concerns the question how to express a dynamical domain in logic without explicitly specifying which conditions are *not* affected by an action. The name "frame problem" derives from a classical technique used by animated cartoon designers called "framing". Put succinctly, the currently moving parts of the cartoon are superimposed on the "frame" depicting the background of the scene that does not change. In logical settings, actions are usually specified by what they change (that is by their results), implicitly assuming that everything else (the frame) remains unchanged. Therefore, we need some way of declaring the general rule-of-thumb that an action is assumed not to change a given property of a situation unless this is stated explicitly. This default assumption is known as the *law of inertia*.

Technically, the frame problem amounts to the challenge of formalizing this law (see for example, Reiter (1991) and Sandewall (1994)). The main obstacle appears to be the monotonicity of classical logic: if a set of premises is extended, then the set of logical conclusions from the extended set includes the conclusions from the original set of premises. Researchers have developed a variety of non-monotonic formalisms, such as circumscription (McCarthy, 1986) and have investigated their application to the frame problem. Alas, none of this turned out to be straightforward. In contrast, in logics of programs such as PDL, the problem disappears because the results of previous actions are naturally preserved in the course of executing subsequent actions in a sequence.

Later, the term "frame problem" acquired a broader meaning in philosophy, where it is formulated as the problem of limiting the beliefs that have to be updated in response to actions.

6.4.1.8 Frame axioms for Plan Generation

In TEAMLOG^{dyn} we assume that the system developer takes care that all complex actions applicable in plan generation, when carried out in the appropriate order, do not disturb the partial planning results obtained previously. For example, the result of division stays intact during subsequent means-end-analysis, action-allocation and construction. In fact, thanks to the dynamic logic framework we have the following general property.

Property

In all Kripke models $\mathcal{M}$ and worlds w, and for all complex actions β_1, β_2, we have:

$$\mathcal{M}, w \models succ(\beta_1; \beta_2) \rightarrow succ(\beta_2)$$

Thus, the following *frame properties* should be preserved:

FR1

$$succ(\text{action-allocation}(\tau, P); \text{construction}(\varphi, G, P))$$

$$\rightarrow allocation(\tau, P)$$

FR2

$$succ(\text{means-end-analysis}(\sigma, \tau); \text{action-allocation}(\tau, P);$$

$$\text{construction}(\varphi, G, P)) \rightarrow means(\sigma, \tau) \wedge allocation(\tau, P)$$

FR3

$$succ(\text{division}(\varphi, \sigma); \text{means-end-analysis}(\sigma, \tau);$$

$$\text{action-allocation}(\tau, P); \text{construction}(\varphi, G, P))$$

$$\rightarrow division(\varphi, \sigma) \wedge means(\sigma, \tau) \wedge allocation(\tau, P)$$

The above properties are necessary to reason about the complex social actions during plan generation. However, they do not exhaust frame axioms required in TEAMLOG reasoning.

6.5   Evolution of Commitments During Reconfiguration

Collective commitment triggers plan realization. Once the process is underway, due to action failure or other circumstances, the collective commitment may evolve to ensure a certain realization of a common goal, if possible. The evolution of collective commitment may imply the evolution of both collective intention and the cooperative team.

Even though our understanding of collective commitment is intuitively appealing, its complexity calls for a rigorous maintenance of its constituents during teamwork. Our methodological approach is built on the reconfiguration algorithm, regardless the type of applied collective commitment, as agents' awareness about details of the situation is left aside. In our analysis we aim at formulating a *minimal* set of properties ensuring correctness of the reconfiguration.

6.5.1   Commitment Change: Zooming Out

In short, during plan execution a number of different cases is treated by the reconfiguration algorithm, all of them leading to changes in the agents' attitudes. It may be helpful to keep in mind the analogue of backtracking. In the successful case, all agents successfully perform their actions, leading to system-success (φ) (see Case 1, Section 6.5.2.1). Otherwise, the unsuccessful Case 2 is split into a number of subcases (2a, 2b, 2c, 2d and 2e), according to the reasons of failure and the possibility of re-allocating failed actions to other agents. In this situation we speak about new action allocation, followed by the necessary attitudes' revision of the agents involved in the exchange.

- A new action allocation succeeds (case 2a);
- A new action allocation fails; and

  – A failed action blocks achieving the overall goal (case 2b); or
  – No failed action blocks achieving the goal; and
    ▪ a new means-end analysis, followed by action allocation, succeeds (case 2c); or
    ▪ a new means-end analysis, followed by action allocation, fails; and
      ◇ a new task division, followed by means-end analysis and action allocation, succeeds (case 2d); or
      ◇ a new task division, followed by means-end analysis and action allocation, fails (case 2e).

6.5.2   Commitment Change: Case by Case

Now, case by case, a formal property of commitment evolution will be formulated, proved and illustrated in action. In the proofs below, we will make use of two general properties of complex actions, that follow immediately from the correct construction of the function *perfco* and the definition of confirm in dynamic logic (see Definiton 6.6 and Section 6.3.1). In fact, the first property has already been presented in Section 6.4.1, while the second property is just axiom **P3** of dynamic logic (see Section 6.3.1). Here they are gathered together for easy reference.

Lemma 6.1 *In all Kripke models $\mathcal{M}$ and worlds w and for all complex actions β_1, β_2, we have*

$$\mathcal{M}, w \models succ(\beta_1; \beta_2) \rightarrow succ(\beta_2) \tag{6.1}$$

$$\mathcal{M}, w \models [\mathtt{confirm}(\psi)]\chi \leftrightarrow (\psi \rightarrow \chi) \tag{6.2}$$

We will illustrate all cases with a theorem-proving example, a variation of the one introduced in Section 5.4. We give a short reminder here and then formally describe all cases.

Running example Let us consider a system with the goal to prove a certain new mathematical theorem. In this domain, we decided to adopt the conservativity assumption. Moreover we assume that all agents are prepared to work in a distributed environment, communicating, coordinating and negotiating when necessary. Suppose that the system starts with an initiator t (theorem prover) who already formed a team $G = \{t, l, c\}$ that has established a collective intention to overall goal $\varphi =$ "theorem T has been proved".

During task division agents created the sequence of subgoals $\sigma = \langle \sigma_1, \sigma_2 \rangle$, with:

$$\sigma_1 = \text{'lemmas relevant for } T \text{ have been proved'} \text{ and}$$

$$\sigma_2 = \text{'theorem } T \text{ has been proved from lemmas'}.$$

During means-end analysis, complex actions have been found to achieve these subgoals, namely the sequence $\tau = \langle provelemma1, provelemma2, checklemma1, checklemma2, provetheorem, checktheorem \rangle$. During action allocation the team divided these actions among themselves and created a temporal structure, resulting in social plan P. Note that the example is a bit more sophisticated than the one of Section 5.4, in that agents now plan to work in parallel as well as sequentially (first l and t in parallel prove the lemmas and the theorem; then c checks all proofs):

$$P = \langle\langle\langle\langle provelemma1, l\rangle; \langle provelemma2, l\rangle\rangle \parallel \langle provetheorem, t\rangle\rangle;$$

$$\langle\langle\langle checklemma1, c\rangle; \langle checklemma2, c\rangle\rangle; \langle checktheorem, c\rangle\rangle\rangle$$

They agreed that their plan was correct ($constitute(\varphi, P)$) and publicly established pairwise social commitments:

$$\text{COMM}(l, t, provelemma1) \wedge \text{COMM}(l, t, provelemma2) \wedge$$

$$\text{COMM}(t, l, provetheorem) \wedge \text{COMM}(c, l, checklemma1) \wedge$$

$$\text{COMM}(c, l, checklemma2) \wedge \text{COMM}(c, t, checktheorem)$$

6.5.2.1 Case 1: the Successful Case

When everything goes right during team action, all agents successfully executed their actions from plan P.

Property: the successful case
If a collective commitment S-COMM$_{G,P}(\varphi)$ holds and a plan P has just been successfully executed, then φ holds. In other words, for all Kripke models $\mathcal{M}$ in which the teamwork axioms hold, and all worlds w:

$$\mathcal{M}, w \models \text{S-COMM}_{G,P}(\varphi) \rightarrow [\texttt{confirm}(succ(P))]\varphi$$

Proof Suppose $\mathcal{M}, w \models \text{S-COMM}_{G,P}(\varphi)$. Then, using the definition of strong collective commitment, $\mathcal{M}, w \models constitute(\varphi, P)$. Finally, by axiom **CS**:

$$\mathcal{M}, w \models [\texttt{confirm}(succ(P))]\varphi$$

The example In this case, l has proved the two lemmas, t has proved the theorem from these lemmas and c has found all the proofs to be correct. Indeed, after such a successful plan execution, the overall goal has been achieved, that is theorem T has been proved.

6.5.2.2 Case 2: an Action Failed

In the sequel, some actions fail during team action. Then, we show the evolution of collective commitment according to the reasons for failure given in the reconfiguration algorithm. Inevitably, the "old" collective commitment has to be dropped because the social commitments with respect to the failed actions from S-COMM$_{G,P}(\varphi)$ do not exist anymore. After an action failure, the situation is not a priori hopeless: the collective commitment may still evolve, leading to a good end. This evolution is done according to a conservative revision of the social plan P, resulting in a new plan P'. The particular cases will differ with respect to the stage where the re-planning actually starts: at action allocation, at means-end analysis or even earlier, at task division. Thus, we split this situation into four cases of varying difficulty.

Case 2a: Reallocation possible

When other agents are prepared to realize the failed actions, that is, when action re-allocation is possible, a new plan P' is devised, starting from a new action allocation. In this way the results of the previous task division and means-end analysis are conserved, taking minimal costs: only a new action allocation is performed. Finally, a new collective commitment based on P' is constructed. This is expressed by the property below.

Property: reallocation possible

Suppose that there is an $(i, \alpha) \in P$ such that *failed*(i, α) and *objective*$_G(\alpha)$ and no failed α blocks φ, that is, $\neg$*necessary*(α, φ) holds for all objectively failed actions. Then for the current action sequence τ and a new social plan P' we have for all Kripke models $\mathcal{M}$ in which the teamwork axioms hold, and all worlds w:

$$\text{C-INT}_G(\varphi) \wedge division(\varphi, \sigma) \wedge means(\sigma, \tau) \rightarrow$$

$$[\texttt{confirm}(succ(\texttt{action-allocation}(\tau, P'); \texttt{construction}(\varphi, G, P')))]$$

$$\text{S-COMM}_{G,P'}(\varphi)$$

Proof Suppose $\mathcal{M}, w \models \text{C-INT}_G(\varphi) \wedge division(\varphi, \sigma) \wedge means(\sigma, \tau)$. Now by the second property in Lemma 6.1, it suffices to show that if:

$$\mathcal{M}, w \models succ(\texttt{action-allocation}(\tau, P'); \texttt{construction}(\varphi, G, P')),$$

then $\mathcal{M}, w \models \text{S-COMM}_{G,P'}(\varphi)$; so suppose:

$$\mathcal{M}, w \models succ(\texttt{action-allocation}(\tau, P'); \texttt{construction}(\varphi, G, P'))$$

It immediately follows by axiom **FR1** that $\mathcal{M}, w \models allocation(\tau, P')$. Combined with $\mathcal{M}, w \models division(\varphi, \sigma) \wedge means(\sigma, \tau)$ this implies by axiom **C0** that $\mathcal{M}, w \models constitute(\varphi, P')$. On the other hand, by the first property of Lemma 6.1 we derive $\mathcal{M}, w \models succ(\texttt{construction}(\varphi, G, P'))$ from:

$$\mathcal{M}, w \models succ(\texttt{action-allocation}(\tau, P'); \texttt{construction}(\varphi, G, P'))$$

Thus we have:

$$\mathcal{M}, w \models \text{C-INT}_G(\varphi) \wedge constitute(\varphi, P') \wedge succ(\texttt{construction}(\varphi, G, P')$$

and so by postulate **CTR** we conclude $\mathcal{M}, w \models \text{S-COMM}_{G,P'}(\varphi)$, as desired.

The example Suppose that l does not succeed in proving Lemma 1 and in fact believes that it cannot as it misses some knowledge about elliptic curves, which t does have. After t communicates that it will pitch in for l, COMM($l, t, provelemma1$) (and thus the old collective commitment) is dropped, and a new social plan is devised, for example:

$$P = \langle\langle\langle provelemma2, l\rangle \parallel \langle\langle provelemma1, t\rangle; \langle provetheorem, t\rangle\rangle\rangle$$

$$\langle\langle\langle checklemma1, c\rangle; \langle checklemma2, c\rangle\rangle; \langle checktheorem, c\rangle\rangle\rangle$$

Finally, a new strong collective commitment is constructed, containing the social commitment $COMM(t, l, provelemma1)$.

Case 2b: some failed action blocks the goal

In this case some action α that was *necessary* for achieving the goal failed and cannot be re-allocated, that is $objective_G(\alpha)$ and $necessary(\alpha, \varphi)$ hold. This is the most serious negative case, inevitably leading to system-failure.

Property: goal blocked
Suppose that there is an $(i, \alpha) \in P$ such that $failed(i, \alpha)$ and $objective_G(\alpha)$ and α blocks φ, that is, $necessary(\alpha, \varphi)$ holds for an objectively failed action. Then for all Kripke models $\mathcal{M}$ in which the teamwork axioms hold, and all worlds w:

$$\mathcal{M}, w \models failed(i, \alpha) \wedge objective_G(\alpha) \wedge necessary(\alpha, \varphi) \rightarrow$$

$$do(\texttt{system-failure}(\varphi))$$

This formalizes the postulate to be ensured by the system designer. Thus, if it is discovered that a failed action blocks φ, the system fails to achieve φ and stops. This implies that neither a collective intention nor an evolved collective commitment towards φ will be established. In the Appendix, an alternative account of this case is formalized in the language of branching time temporal logic, with a focus on formalizing the concept of blocking.

The example Suppose that, while checking t's proof of the theorem from the lemmas, c discovers that not only the proof is wrong but also finds a counterexample to the theorem. Then nothing can be done to remedy the problem. This concludes the case.

Case 2c: New means-end analysis possible

In this case action reallocation is not possible because there are some objectively failed actions. This means that for every relevant social plan P', allocation with respect to the current action sequence τ fails. Furthermore, in this case each objectively failed action does not block the goal. In this situation, the old collective commitment is dropped but its evolution is still possible, if a new means-end analysis yields new actions realizing the failed subgoals, allowing for a new allocation of them. This is expressed by the following property.

Property: new means-end analysis possible
Suppose that there is an $(i, \alpha) \in P$ such that $failed(i, \alpha)$ and $objective_G(\alpha)$ and no failed α blocks φ, that is, $\neg necessary(\alpha, \varphi)$ holds for all objectively failed actions. Then for the current goal sequence σ and action sequence τ and for every social plan P', there are

τ' and P'' excluding the objectively failed actions such that the following holds for all Kripke models $\mathcal{M}$ in which the teamwork axioms hold, and all worlds w:

$$\text{C-INT}_G(\varphi) \wedge division(\varphi, \sigma) \rightarrow$$

$$[\text{confirm}(failed(\text{action-allocation}(\tau, P')))]$$

$$[\text{confirm}(succ(\text{means-end-analysis}(\sigma, \tau');$$

$$\text{action-allocation}(\tau', P''); \text{construction}(\varphi, G, P'')))]$$

$$\text{S-COMM}_{G, P''}(\varphi)$$

Proof Suppose $\mathcal{M}, w \models \text{C-INT}_G(\varphi) \wedge division(\varphi, \sigma)$. Now by the second property in Lemma 6.1, it suffices to show that if:

$$\mathcal{M}, w \models succ(\text{means-end-analysis}(\sigma, \tau'); \text{action-allocation}(\tau', P'');$$

$$\text{construction}(\varphi, G, P''));$$

then $\mathcal{M}, w \models \text{S-COMM}_{G, P''}(\varphi)$; so suppose:

$$\mathcal{M}, w \models succ(\text{means-end-analysis}(\sigma, \tau'); \text{action-allocation}(\tau', P'');$$

$$\text{construction}(\varphi, G, P'')).$$

It immediately follows by axiom **FR2** that $\mathcal{M}, w \models means(\sigma, \tau') \wedge allocation(\tau', P'')$. Combined with $\mathcal{M}, w \models division(\varphi, \sigma)$ this implies by axiom **C0** that $\mathcal{M}, w \models constitute(\varphi, P'')$.

On the other hand, by the first property of Lemma 6.1 we derive:

$$\mathcal{M}, w \models succ(\text{construction}(\varphi, G, P''))$$

and, exactly as in case 2a, we derive $\mathcal{M}, w \models \text{S-COMM}_{G, P''}(\varphi)$ by **CTR**.

The example As in case 2a, suppose that l does not succeed in proving Lemma 1, but now t and c do not believe they can prove it, either. The team does a new means-end analysis based on the old subgoal sequence, and comes up with some other lemmas (say 3, 4 and 5) that together hopefully imply the theorem. This gives rise to a new action sequence $\tau' = \langle provelemma3, provelemma4, provelemma5, checklemma3, checklemma4, checklemma5, provetheorem, checktheorem \rangle$. They allocate the actions in a similar way as before, creating a new social plan P'', for example:

$$P'' = \langle\langle\langle\langle\langle provelemma3, l\rangle; \langle provelemma4, l\rangle\rangle;$$

$$\langle provelemma5, l\rangle\rangle \parallel \langle provetheorem, t\rangle\rangle;$$

$$\langle\langle\langle\langle checklemma3, c\rangle; \langle checklemma4, c\rangle\rangle;$$

$$\langle checklemma5, c\rangle\rangle; \langle checktheorem, c\rangle\rangle\rangle.$$

Finally, by public communication they establish new social commitments leading to a new strong collective commitment.

Case 2d: New task division possible

When no objectively failed action blocks the goal but neither action reallocation, nor a new means-end analysis is possible for the failed actions, this means that for the current τ, action allocation fails to deliver any social plan P' and then means-end analysis with respect to the current σ fails to deliver any action sequence τ' not containing the objectively failed actions. Even in this difficult case, the evolution of collective commitment is still possible. This happens when task division for the goal φ is successfully executed, resulting in a new goal sequence σ'. Then, this sequence is a subject of a new round of means-end analysis, establishing a new action sequence τ''. Next follows action allocation, to create a new social plan P'' on the basis of τ''. The following property describes the result.

Property: new task division possible

Suppose there is an $(i, \alpha) \in P$ such that *failed* (i, α) and for all failed α, $\neg necessary\,(\alpha, \varphi)$. Then for the current goal sequence σ and action sequence τ, and for every social plan P' and action sequence τ', there are σ', τ'' and P'' such that:

$$\text{C-INT}_G(\varphi) \rightarrow$$

$$[\text{confirm}(failed(\text{action-allocation}(\tau, P')))]$$

$$[\text{confirm}(failed(\text{means-end-analysis}(\sigma, \tau')))]$$

$$[\text{confirm}(succ(\text{division}(\varphi, \sigma'), \text{means-end-analysis}(\sigma', \tau'');$$

$$\text{action-allocation}(\tau'', P''); \text{construction}(\varphi, G, P'')))]$$

$$\text{S-COMM}_{G, P''}(\varphi)$$

Proof Suppose $\mathcal{M}, w \models \text{C-INT}_G(\varphi)$. By the second property in Lemma 6.1, it suffices to show that if:

$$\mathcal{M}, w \models succ(\text{division}(\varphi, \sigma'); \text{means-end-analysis}(\sigma', \tau'');$$

$$\text{action-allocation}(\tau'', P''); \text{construction}(\varphi, G, P''));$$

then $\mathcal{M}, w \models \text{S-COMM}_{G, P''}(\varphi)$; so suppose:

$$\mathcal{M}, w \models succ(\text{division}(\varphi, \sigma'); \text{means-end-analysis}(\sigma', \tau'');$$

$$\text{action-allocation}(\tau'', P''); \text{construction}(\varphi, G, P'')).$$

It immediately follows by axiom **FR3** that:

$$\mathcal{M}, w \models division\,(\varphi, \sigma') \wedge means\,(\sigma', \tau'') \wedge allocation\,(\tau'', P'')$$

This implies by axiom **C0** that $\mathcal{M}, w \models constitute\,(\varphi, P'')$.
On the other hand, by the first property of Lemma 6.1 we derive:

$$\mathcal{M}, w \models succ(construction\,(\varphi, G, P''))$$

and, exactly as in case 2a, we conclude $\mathcal{M}, w \models \text{S-COMM}_{G, P''}(\varphi)$ by **CTR**.

The example Suppose that the theorem has been divided into lemmas several times and each time it was impossible to prove some essential lemma. Then the team concludes that they are not able to prove the theorem by formulating and proving suitable lemmas. Then they may come up with a completely different task division, for example $\sigma' = \langle \sigma_3, \sigma_4 \rangle$ where $\sigma_3 =$ "a theorem analogous to T has been found in a different area of mathematics" and $\sigma_4 =$ "a suitable translation between the two contexts has been defined". On means-end analysis and action allocation result in a social plan P'' very different from P.

In case 2d, if task division is not successful, the story of the current team is completed and a return to team formation is made in order to establish a new team attempting to achieve φ. In this way, the evolution of the collective commitment is completed as well.

6.5.3 Persistence of Collective Intention

During the evolution of collective commitment within a fixed team, the agents could exchange their individual actions and create new social plans, as long as the group was consolidated through a collective intention.

The problem of persistence of individual and collective motivational attitudes calls for a careful coordination of the agent's personal and team perspective. For example, when an agent succeeds in its action, it inevitably drops the corresponding social commitment. On the other hand, it remains involved in the team effort regarding:

- its own social commitment(s) towards other actions;
- monitoring the agents who have committed to it;
- its awareness about, and plan correctness;
- the underlying collective intention.

Let us recall that when a collective intention no longer exists, the group may disintegrate. Therefore, the individual agents carry a special responsibility to protect the collective intention and thus to refrain from dropping their corresponding individual intention if it is not absolutely necessary.

The persistence of collective intention is a necessary condition for collective commitment to hold. On the other hand, due to dynamic circumstances, social commitments may naturally change according to agents' decisions, based on their individual commitment strategies. If these possibilities are exploited but the team cannot work for the common goal anymore, the team must disintegrate. Then, the old collective intention is dropped, leading to the demise of the associated collective commitment. According to the reconfiguration algorithm, a new team is created, a collective intention towards the goal φ within this team is established and in this way *plan formation* starts again.

6.6 TEAMLOG Summary

In the research presented in this chapter, static TEAMLOG notions are confronted with the dynamics of teamwork in a changeable and unpredictable environment. As before, the resulting properties of TEAMLOG^{dyn} express solely vital aspects of teamwork, leaving room for case-specific extensions. Within this scope, both the static and the dynamic part of the theory yield a set of *teamwork axioms*. They constitute both a definition of motivational attitudes in BDI systems and a specification of their dynamic evolution.

In the MAS literature (see for example Tambe, 1996) some phenomena such as the dynamics of attitude revision during reconfiguration have barely received prior attention. In this chapter, we fill this gap. Our notion of collective commitment ensures efficiency of reconfiguration in two ways. Unlike in Wooldridge and Jennings (1999), our approach to group commitments is formalized in a non-recursive way. This allows for a straightforward revision. Next, because only social commitments to individual actions appear, it often suffices to revise just some of them. In that way we avoid involving the whole team in re-planning. Such an approach has pragmatic power: agents can take the whole process of building, updating and revising collective commitments into their own hands. Relevant aspects have been first treated in Chapter 5 and then formally proved to be correct in this chapter. Thus, teamwork axioms may serve a system designer as a high-level specification at design-time. During run-time, formal verification methods may be applied to check the correctness of the system behavior.

Let us stress the novelty of using dynamic logic to express collective attitude dynamics in BDI systems. The language of dynamic logic allows us to precisely formulate both the preconditions and the results of complex social actions during reconfiguration. However, the framework of normal modal logics we apply is based on standard Kripke semantics and so like other similar modal logics, it suffers from the well-known logical omniscience problem as discussed earlier. Because of the necessitation rule, agents are supposed to know and intend all tautologies; also, because of the distribution axiom, they are supposed to know all logical consequences of their knowledge and to intend all logical consequences of their intentions. This is clearly unrealistic. For epistemic logic, several solutions to the logical omniscience problem have been proposed, mostly based on non-normal modal logics (see Chapter 2). Similar solutions have been proposed for individual intentions (see Konolige and Pollack, 1993). The question how to design a non-normal multi-modal logic suitable to solve logical omniscience problems in TEAMLOG still remains open.

Grant *et al.* (2005a) provide an interesting comparison of six different approaches to teamwork, called by them as follows: the *Joint Intentions approach* (Levesque *et al.*, 1990), the *Team Plans* approach (Sonenberg *et al.*, 1992), the *SharedPlans* approach (Grosz and Kraus, 1996, 1999), the *CPS* approach (Wooldridge and Jennings, 1999), our Collective Intentions approach Dunin-Kȩplicz and Verbrugge, 2002 and the Cooperative Sub-contracting approach (Grant *et al.*, 2005b). Grant *et al.* (2005a) introduce an example task that is quite easy but still requires cooperation between at least two agents. They need to go to a location, where a large and heavy block lies, which needs to be pushed to a new location, while avoiding an obstacle. The example is formalized in all six approaches, highlighting the special focus of each approach and pointing to advantages and disadvantages of each. Then, they evaluate the six frameworks against Bratman's four criteria for shared cooperative activity (Bratman, 1992):

- mutual responsiveness (for example a musician hearing and responding to the notes of his/her colleague);
- commitment to the joint activity;

Table 6.4 Summary of evaluations from Grant *et al.* (2005a).

	Joint intentions	Team plans	SharedPlans	CPS	Collective intentions	Cooperative subcontracting
Authors	Cohen and Levesque	Sonenberg group	Grosz and Kraus	Wooldridge and Jennings	Dunin-Keeplicz and Verbrugge	Grant, Kraus and Perlis
Focus	Persistence of joint intentions	Team formation, role assignment, establishment of joint goals, action complexity	Individual intentions and actions in teamwork, plan and sub-plan execution and coordination in explicit time structure	Joint commitment to a goal, potential for cooperation, mental states based on a branching-time model	Mental attitudes in cooperation, formal modeling of group consistency, failure recovery	Subcontracting and task sharing
Formalism	Modal logic and Kripke models	Modal logic and Kripke models	Syntactic description	Modal logic and Kripke models	Modal logic and Kripke models	Syntactic description
Emphasis on single or multi-agent point of view	Multi-agent mental attitudes	Single-agent mental attitudes leading to emergent group attitudes	Single-agent mental attitudes leading to emergent group attitudes	Multi-agent mental attitudes	Multi-agent mental attitudes	Single-agent mental attitudes leading to emergent group attitudes
Time representation	Explicit representation in plans	Time-tree in the semantics	Explicit representation in plans	Paths in a branching-time structure	Doesn't deal with time (implicit)	Explicit representation in plans
Support for expression of complex plans	No	Yes, great emphasis	Yes, specific constructs	No	No	Yes, specific constructs

Coverage of four stages of cooperative behavior	No, deals only with establishment of group intentions	Yes	Yes	Yes	Yes	Yes
Bratman's criteria of 'shared cooperative activity'	Commitment to joint activity, mutual support and mutual responsiveness	Commitment to joint activity, meshing subplans, mutual responsiveness	All criteria handled explicitly	Commitment to joint activity, implicit meshing of subplans and mutual support in plan formation process	Commitment to joint activity, mutual responsiveness and mutual support are explicit via mental attitudes, meshing of subplans is implicit	Commitment to joint activity, mutual responsiveness, meshing of subplans in the execution phase

- commitment to mutual support;
- formation of subplans that mesh with one another.

In Table 6.4, we summarize the evaluations from Grant *et al.*, (2005a), changing the evaluation of "support for the expression of complex plans" in our approach to "yes", according to the dynamic framework TEAMLOG^{dyn}, while Grant *et al.*, (2005a)'s "no" applies solely to the static TEAMLOG of Dunin-Kęplicz and Verbrugge (2002).

7

A Case Study in Environmental Disaster Management

Do you want to improve the world?
I don't think it can be done.

Tao Te Ching (Lao-Tzu, Verse 29)

7.1 A Bridge from Theory to Practice

Disaster management is a broad discipline related to dealing with and avoiding risks (Wisner *et al.*, 2004). This involves several important tasks: *preparing for disaster* before it occurs, *disaster response* (for example emergency evacuation and decontamination) and *restoration* after natural or human-made disasters have occurred. In general, disaster management is the continuous process by which all individuals, groups and communities manage hazards in an effort to avoid or ameliorate the impact of disasters resulting from them. Actions taken depend in part on perceptions of risk of those exposed (Cuny, 1983). Activities at each level (individual, group, community) affect the other levels.

In this chapter we focus on *disaster response* and, more specifically, on *decontamination* of a certain polluted area (Dunin-Kęplicz *et al.*, 2009b). We show how to make a bridge between theoretical foundations of a BDI system and their application. The case study presents the interaction and cooperation between agents, outlines their goals and establishes the necessary distribution of knowledge and commitment throughout the team. Importantly, we show how to tune TEAMLOG to the application in question by establishing sufficient, but still minimal levels of team attitudes.

In TEAMLOG, the main subject of tuning is *awareness* of individuals and teams. As indicated before, group awareness is usually expressed in terms of *common belief*, fully reflecting collective aspects of agents' behavior. Due to its infinitary flavor, this concept has a high complexity: its satisfiability problem is EXPTIME-complete (see Chapter 9).

Teamwork in Multi-Agent Systems: A Formal Approach   Barbara Dunin-Kęplicz and Rineke Verbrugge
© 2010 John Wiley & Sons, Ltd

There are some general ways to reduce the complexity: by restricting the language, by allowing only a small set of atomic propositions or restricting the modal context in formulas, as proved in Chapter 9 and Dziubiński (2007). Apart from these methods, when building a specific multi-agent system, the use of domain knowledge is crucial in tailoring TEAMLOG to the circumstances in question. In the case study about the prevention of ecological disasters we will show how to adjust the infinitary notions of collective attitudes to a real-world situation. This can be achieved by applying weak forms of awareness which essentially reduce the complexity of team attitudes.

This chapter, based on joint work with Michał Ślizak (Dunin-Kęplicz *et al.*, 2009b), is structured as follows. In Section 7.2, some definitions and assumptions regarding the environment are presented, including an outline of the interactions within and between teams. This is followed in Section 7.3 by definitions of social plans. In Section 7.4 we explore the minimal requirements for successful teamwork in environmental disaster response, which is summed up by a short discussion.

7.2   The Case Study: Ecological Disasters

The case study deals with ecological disasters caused by specific poisons. Their prevention and repair will be performed by means of heterogeneous multi-agent teams, which are applicable in situations where time is critical and resources are bounded (Kleiner *et al.*, 2006; Sycara and Lewis, 2004). The maintenance goal *safe* is to keep a given region *REG* safe or to return it to safety if it is in danger.

Possible hazards are two kinds of poison, X_1 and X_2, which are dangerous in high concentrations. They may be explosive if they react with one another to form compound $X_1 \oplus X_2$, which happens at high concentrations. Three functions f_1, f_2 and f_3 reflect the influence of temperature $t(A)$, pressure $p(A)$ and concentrations $c_1(A)$ and $c_2(A)$ of poisons X_1 and X_2 at location A on the possible danger level at that location. The function ranges are divided into three intervals, as follows:

The first poison X_1:
- *safe*$_1$ iff $f_1(p(A), t(A), c_1(A)) \in [0, v_1]$;
- *risky*$_1$ iff $f_1(p(A), t(A), c_1(A)) \in (v_1, n_1]$;
- *dangerous*$_1$ iff $f_1(p(A), t(A), c_1(A)) \in (n_1, \infty)$.

The second poison X_2:
- *safe*$_2$ iff $f_2(p(A), t(A), c_2(A)) \in [0, v_2]$;
- *risky*$_2$ iff $f_2(p(A), t(A), c_2(A)) \in (v_2, n_2]$;
- *dangerous*$_2$ iff $f_2(p(A), t(A), c_2(A)) \in (n_2, \infty)$.

The compound poison $X_1 \oplus X_2$:
- *safe*$_3$ iff $f_3(p(A), t(A), c_1(A), c_2(A)) \in [0, v_3]$;
- *risky*$_3$ iff $f_3(p(A), t(A), c_1(A), c_2(A)) \in (v_3, n_3]$;
- *explosive* iff $f_3(p(A), t(A), c_1(A), c_2(A)) \in (n_3, \infty)$.

We define *safe* := *safe*$_1 \wedge$ *safe*$_2 \wedge$ *safe*$_3$ and refer to it as a goal and as a predicate. There are also relevance thresholds ε_1 and ε_2: when the concentration of a poison X_i exceeds ε_i, the respective function f_i is computed.

7.2.1 Starting Point: the Agents

This model reflects cooperation between humans, software agents, robots and unmanned aerial vehicles (*UAVs*), as discussed in Doherty *et al.* (2006) and WITAS (2001), and a helicopter steered by a pilot.

The whole process is coordinated by one *coordinator*, who initiates cooperation, coordinates teamwork between different subteams of the full team G, is responsible for dividing the disaster zone into sectors and assigning a subteam to each sector to perform clean-up. Several subteams $G_1, \ldots G_k \subseteq G$ of similar make-up work in parallel, aiming to prevent or neutralize a contamination. Each of these subteams G_i consists of:

- One UAV_i – responsible to the coordinator for keeping assigned sectors in a safe state. This agent cannot carry a heavy load, but can carry the computer and therefore has considerable computational capabilities for planning and is capable of observing and mapping the terrain.
- n_i identical neutralizing robots $rob_{i_1}, \ldots, rob_{i_{n_i}}$ – responsible to their UAV_i for cleaning up a zone.

In addition to the subteams, there is also a rather independent member of the team G:

- One regular helicopter steered by the human *pilot*, who can independently choose the order of cleaning up assigned areas is directly accountable to the coordinator and can communicate as equals with the *UAVs*.

See Figure 7.1 for the team structure.

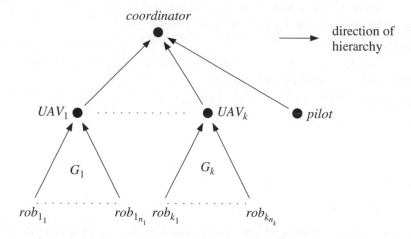

Figure 7.1 Hierarchical team structure of the ecological disaster prevention and repair team G.

7.2.2 Cooperation between Subteams

The entire disaster zone is divided into sectors by the coordinator, based on terrain type, subteam size and hot spots known in advance. Subteams are responsible for (possibly many) sectors. The leader UAV_i of a subteam G_i prepares a plan P_i to keep its sectors safe. Each plan is judged based on a fitting function *fit*, which takes into account:

- available robots, including their current task, load, capacity and position;
- whether the plan relies on the help from other subteams;
- task priorities;
- the minimum amount of time it takes to implement;
- the minimum number of robots it requires.

The *UAVs* communicate and cooperate with one another. If performing tasks requires more robots than are currently available in their own subteam G_i, its leader UAV_i can call for reinforcements from another UAV_j, for $j \leq k, j \neq i$. Of course for UAV_j in question, fulfilling its own subteam G_j's objectives has priority over helping others from G_i.

7.2.3 A Bird's-Eye View on Cases

To maintain the goal *safe*, the situation is monitored by the coordinator and the *UAVs* on a regular basis, with frequency *freq*. During *situation recognition*, in the risky cases monitoring is performed twice as frequently. Depending on the mixture and density of poisons in a location, some general cases followed by the relevant procedures are established. All remedial actions are to be performed relative to the contaminated area:

Case *safe*:
   true $\longrightarrow$ *situation recognition*

Case *dangerous*$_1$:
   rain $\longrightarrow$ *liquid L_1 to be poured on the soil*
   normal or dry $\longrightarrow$ *liquid L_2 to be sprayed from the air*

Case *dangerous*$_2$:
   rain $\longrightarrow$ *solid S_1 to be spread*, followed by *liquid catalyst K_1 to be poured*
   normal or dry $\longrightarrow$ *solid S_1 to be spread*

Case *explosive*:
   before explosion $\longrightarrow$ *evacuation*
   after explosion $\longrightarrow$ *rescue action*

In the next section, we delineate some of these global plans. We do not present too many details of the plans, nor do we discuss failure handling.

7.3 Global Plans

In order to control the amount of interactions and decrease the time needed to establish beliefs, the applied team model is hierarchical. The coordinator views a subteam G_i as a single cleaning agent, even though the *UAVs* manage the work of many autonomous neutralizing robots.

7.3.1 The Global Social Plan ⟨Cleanup⟩

The global social plan for which the coordinator and *UAVs* are responsible, is designed with regard to location A. It is a loop, in which observation is interleaved with treatment of current dangers by decreasing the level of priority, from most to least dangerous. The goal (denoted as *Clean*) is to keep locations in a *safe* state. All subplans mentioned in $\langle Cleanup \rangle$, namely $\langle$ Plan $SR \rangle$, $\langle$ Plan $E \rangle$, $\langle$ Plan $D_1 R \rangle$, $\langle$ Plan $D_1 N \rangle$, $\langle$ Plan $D_2 R \rangle$ and $\langle$ Plan $D_2 N \rangle$, are described more precisely in the subsequent subsections.

```
begin
  freq := a; {freq - interval between two checks of the environment}
  loop
    ⟨ Plan SR⟩ {Compute the situation at A, with frequency freq}
    if explosive then do ⟨ Plan E⟩ end;
    elif dangerous₁ and rain then do ⟨ Plan D₁R⟩ end;
    elif dangerous₁ then do ⟨ Plan D₁N⟩ end;
    elif dangerous₂ and rain then do ⟨ Plan D₂R⟩ end;
    elif dangerous₂ then do ⟨ Plan D₂N⟩ end;
    elif risky₁ ∨ risky₂ ∨ risky₃ then freq:= a/2 end
    else {safe situation} freq := a end;
  end
end.
```

Here, a represents the frequency with which the environment should be checked: this interval is shortened when a risky situation is encountered.

7.3.2 The Social Plan $\langle SR \rangle$

This plan performs situation recognition at location A. One of the *UAVs*, for example UAV_1, is responsible for monitoring. Alternatively, situation recognition could be assigned as a joint responsibility to $UAV_1, \ldots, UAV_k$; however, that solution would require information fusion which is in general a very complex process.

```
begin
  C₁ := c₁(A) {C₁ is the measured concentration of poison X₁ at A}
  C₂ := c₂(A) {C₂ is the measured concentration of poison X₂ at A}
  T := t(A) {T is the measured temperature at A}
  P := p(A) {P is the measured air pressure at A}
  {Computation of the situation at A}
  if C₁ > ε₁ then compute f₁(C₁, T, P) end;
  if C₂ > ε₂ then compute f₂(C₂, T, P) end;
  if C₁ > ε₁ and C₂ > ε₂ then compute f₃(C₁, C₂, T, P) end;
end.
```

7.3.3 The Social Plan $\langle E \rangle$

After an explosion, evacuation and rescue of people should take place. This subject is discussed in many studies (Kleiner *et al.*, 2006; Sycara and Lewis, 2004) and will not be elaborated here. Instead, here follow the other subplans included in $\langle Cleanup \rangle$. In these subplans, we assume that the agents start from the base B where neutralizers are stored.

7.3.4   The Social Plan $\langle D_1 R \rangle$

This plan is applicable when $dangerous_1$ occurs under weather condition $rain$. Each UAV_i may be allocated this social plan for a given location as decided by the $coordinator$. Goal $\psi_1(L_1)$ is to apply liquid L_1 on all areas contaminated with poison X_1.

{Assumption: One portion of L_1 neutralizes poison X_1 at a single location.}
```
while contaminated-area ≠ emptyset do
begin
```
$\quad A := calculate(UAV_i, \{rob_{i_j}\})$;
$\quad${UAV_i finds region A for rob_{i_j} to clean up}
$\quad get(rob_{i_j},\ L_1,\ B)$; {$rob_{i_j}$ retrieves a tank with liquid L_1 from location B}
$\quad path := get\_path(UAV_i, rob_{i_j}, B, A)$; {$rob_{i_j}$ requests a path to follow}
$\quad move(rob_{i_j}, path)$; {$rob_{i_j}$ moves from location B to location A}
$\quad pour(rob_{i_j}, L_1, A)$;
```
   contaminated-area := contaminated-area \ A;
```
$\quad return\_path := get\_path(UAV_i, rob_{i_j}, A, B)$;
$\quad move\ (rob_{i_j}, return\_path)$;
```
end.
```

7.3.5   The Social Plan $\langle D_1 N \rangle$

This plan is applicable when $dangerous_1$ occurs under weather condition $normal$ or dry. The spraying is usually performed by the pilot on request from one of the $UAVs$. In the plan below, UAV stands for any of $UAV_1, \ldots, UAV_k$.

Goal $\psi_2(L_2)$: to spray liquid L_2 on areas contaminated with poison X_1.

{Assumption: One portion of L_2 neutralizes poison X_1 at a single location.}
{Assumption: The helicopter can transport k portions of liquid L_2.}
```
while contaminated-area ≠ emptyset do
begin
```
$\quad request(UAV, coordinator, pilot, \psi(L_2))$;
$\quad confirm(pilot, UAV, coordinator, \psi(L_2))$;
$\quad request(pilot, UAV, list_1, k)$;
$\quad send(UAV, pilot, list_1)$; {$list_1$ has at most k contaminated areas}
$\quad upload(helicopter, L_2)$; {`pilot` retrieves liquid L_2}
```
   take-off (helicopter, B); {pilot takes off from location B}
   do 〈 plan-for-spraying (helicopter, L₂,l)〉;
```
$\quad\quad$ {`pilot` sprays L_2 using his own invented plan}
$\quad confirm(pilot, UAV, done(plan-for-spraying(helicopter, L_2, l))$;
$\quad contaminated-area := contaminated-area \setminus list_1$;
$\quad landing(helicopter, B)$;
$\quad free(pilot, coordinator)$;
```
end.
```

7.3.6 The Social Plan $\langle D_2R \rangle$

This plan is applicable when $dangerous_2$ occurs under weather condition *rain*. Goal $\psi_3(S_1, K_1)$: to spread solid S_1 on all areas contaminated with poison X_2, followed by applying catalyst K_1 to all areas where S_1 is present.

```
{Assumption: One portion each of S₁ and K₁ neutralize poison X₂ at a
single location.}
while contaminated − area ≠ emptyset do
begin
    A := calculate(UAVᵢ, {robᵢⱼ, robᵢₗ});
    {UAVᵢ finds region for robᵢⱼ and robᵢₗ to spread solid and catalyst,
            respectively.}
begin_parallel {two main operations are done in parallel:
    applying a solid to the area, and pouring a catalyst on it}
    {a plan similar to ⟨D₁R⟩, but using S₁:}
    get(robᵢⱼ, S₁, B);  {robᵢⱼ retrieves a portion of solid S₁ from location B}
    path := get_path(UAVᵢ, robᵢⱼ, B, A);  {robᵢⱼ requests a path to follow}
    move(robᵢⱼ, path);  {robᵢⱼ moves from location B to location A}
    pour(robᵢⱼ, S₁, A);
    contaminated-area := contaminated-area \ A;
    return_path := get_path(UAVᵢ, robᵢⱼ, A, B);
    move (robᵢⱼ, return_path);
||
    wait_for(transporting(robᵢⱼ, S₁, A)); {robᵢₗ waits until robᵢⱼ is on the way to A}
    get(robᵢₗ, K₁, B);
    path := get_path(UAVᵢ, robᵢₗ, B, A);
    move(robᵢₗ, path);
    wait_for(spread(S₁, A));  {robᵢₗ waits for someone to spread S₁ in A}
    pour(robᵢₗ, K₁, A);
    return_path := get_path(UAV, robᵢₗ, A, B);
    move(robᵢₗ, return_path);
end_parallel
contaminated-area := contaminated-area \ A;
end.
```

7.3.7 The Social Plan $\langle D_2N \rangle$

This plan is applicable when $dangerous_2$ occurs under weather condition *normal or dry*. Each UAV_i may be allocated this social plan for a given location as decided by the *coordinator*.

Goal $\psi_1(S_1)$ is to apply solid S_1 on all areas contaminated with poison X_2.

```
{Assumption: One portion of solid S₁ neutralizes poison X₂ at a single
 location.}
while contaminated-area ≠ emptyset do
begin
    A := calculate(UAVᵢ, {robᵢⱼ});  {UAVᵢ finds region A for robᵢⱼ to clean up}
    get(robᵢⱼ, S₁, B);  {robᵢⱼ retrieves a portion of solid S₁ from location B}
```

$$path := get\_path(UAV_i, rob_{i_j}, B, A); \quad \{rob_{i_j} \text{ requests a path to follow}\}$$
$$move(rob_{i_j}, path); \quad \{rob_{i_j} \text{ moves from location } B \text{ to location } A\}$$
$$pour(rob_{i_j}, S_1, A);$$
$$contaminated\text{-}area := contaminated\text{-}area \setminus A;$$
$$return\_path := get\_path(UAV_i, rob_{i_j}, A, B);$$
$$move(rob_{i_j}, return\_path);$$
end.

7.4 Adjusting the TEAMLOG Definitions to the Case Study

It does not suffice for agents to have an individual intention to their projection of the social plan. They would still act as individuals, so if something new appears in their region or the circumstances change, calling for re-planning, the group would be helpless to adapt, as not being formed properly. Thus, group attitudes such as *collective intentions* become necessary even in this simple, one would think, situation.

Why is a collective intention and a social plan still not enough to start team action in the case study? Because agents may not feel *responsible* for their share. Thus, they need to commit to performing their part of the plan.

7.4.1 Projections

Before continuing, we need to explain the concept of projections. A plan is written for roles to be adapted by agents in our systems. Each role has requirements, assuring that a robot cannot assume the role of a pilot since it is not capable of flying a helicopter. By a plan projection for agent i, we mean a plan where some of the roles have been assumed by agent i. Similarly, a goal projection for i is the subset of overall goals that i is personally interested in achieving.

In this example, agents naturally commit to their controlling *UAV* which acts as a 'middle manager' on behalf of the *coordinator*. Each *UAV* is committed to the *coordinator* with regard to the task of keeping assigned regions in a *safe* state. The *coordinator* and *UAVs* collectively believe that successfully executing plan $\langle Cleanup \rangle$ in an area leads to the achievement of the *safe* state in that area.

Each agent has its own projection of the overall plan.

- The *coordinator* is aware of the $\langle Cleanup \rangle$ plan in the context of all regions, with each subteam represented by an *UAV*.
- *UAVs* need a projection of the $\langle Cleanup \rangle$ plan in all areas to which they are assigned.
- Robots need to have a projection of the $\langle Cleanup \rangle$ plan only regarding actions which they may take part in.

Now, what is the type of collective commitment fitting to the scenario?

On the subteam level, the UAV_i has the highest level of awareness within its subteam G_i as it knows the entire social plan $\langle Cleanup \rangle$ for a particular region. There is no need for other agents to know the plan's details.

The robots from a subteam G_i need a quite limited awareness of the plan. For example, they need to know the partially instantiated subplans applicable in dangerous situations ($\langle D_1 R \rangle$, $\langle D_2 R \rangle$ or $\langle D_2 N \rangle$). In the relevant weather conditions, they may need to carry out one of these subplans for a specific region previously assigned by the leader UAV_i, who also assigns a specific role to each of the robots in G_i. However, in case the system developer wants to foster a type of teamwork where the robots voluntarily help one another, they will also need to be aware of the subplans assigned to nearby robots. Then they can pitch in for a role that one of its colleagues fails to perform.

With regard to the $\langle Cleanup \rangle$ plan, this corresponds to *weak collective commitment* for subteams.

As a reminder, the subteam knows the overall goal, but not the details of the plan: there is no collective awareness of the plan's correctness, even though there is a global awareness that things are under control:

$$\text{W-COMM}_{G,P}(\varphi) \leftrightarrow \text{C-INT}_G(\varphi) \wedge \textit{constitute}\,(\varphi, P) \wedge \bigwedge_{\alpha \in P} \bigvee_{i,j \in G} \text{COMM}(i, j, \alpha) \wedge$$

$$\text{C-BEL}_G(\bigwedge_{\alpha \in P} \bigvee_{i,j \in G} \text{COMM}(i, j, \alpha))$$

On the team G level, the *coordinator* has the highest awareness. The $UAVs$ mainly need to know their projection of the overall plan and they need to believe that the entire plan has been shared among $UAVs$. The *coordinator* knows both the plan and action allocation. With regard to the $\langle Cleanup \rangle$ plan, this corresponds to *weak collective commitment* on the team level.

7.4.2 Organization Structure: Who is Socially Committed to Whom?

Commitments in the team follow the organizational structure of Figure 7.1. The coordinator is socially committed to achieving the overall goal, by means of the main social plan. He is committed to itself or to the relevant control authority, for example, the national environmental agency for which it works.

Each UAV_i for $i = 1, \ldots, k$ is committed to the coordinator to achieve its part of the plan, namely keeping specified regions in safety.

The robots in G_i for $i = 1, \ldots, k$ commit to perform their share to their leading UAV_i, which has the power to uncommit them. There is a clear hierarchy where the coordinator is the leader of the team G, while the $UAVs$ are 'middle-managers' of subteams. The $UAVs$ sometimes commit to a colleague UAV when some of their robots are temporarily delegated to the other's subteam.

The human pilot has a somewhat special role in that he/she does not manage any subteam. Instead, he/she directly commits to the coordinator, or to $UAVs$ if they request his/her assistance.

7.4.3 Minimal Levels of Group Intention and Awareness

What are the minimal levels of awareness and group intention needed for the agents on subteam and team levels?

The robots – two cases are applicable

1. They act only individually; this is the most limited (and economical) case.
2. They perform a limited form of cooperation, for example, they work together to clean up areas faster, or pitch in for other robots when they are unable to perform their share of labour.

We will consider both cases separately while investigating group attitudes of different types of agents involved in achieving a maintenance goal to keep the region safe.

The level of intention

- In case 1, to act individually, all robots need an individual intention to a common goal. Thus, a general intention E-INT$_G$ is created and suffices.
- In case 2, E-INT$^2_{\{i,j\}}$ will be enough to allow forming two-robot teams that are not competitive internally (but see Section 3.6.3 for a discussion that a two-level intention is not sufficient to preclude competition among two-agent coalitions). If agents are supposed to be strictly cooperative, a two-level definition is in general sufficient for larger teams: all agents intend to achieve the goal in cooperation with the others included in their team.

The level of belief

- In case 1, to act individually each robot i needs an individual belief about every group intention (BEL$(i,$ E-INT$_G(\varphi)))$. This way a general belief E-BEL$_G($E-INT$_G(\varphi))$ is in place and suffices. Moreover, each robot should believe that the distribution of the labour by means of bilateral commitment is done properly. Hence, (E-BEL$_G(\bigwedge_{\alpha \in P} \bigvee_{i,j \in G}$ COMM$(i, j, \alpha))$ is in place. This will allow a potential deliberation about actions. It may also prevent robots from doing all the work by themselves.
- In case 2, E-BEL^2_G will be enough to allow deliberation about other robots' intentions and beliefs (especially E-BEL$^2_G($E-INT$^2_G(\varphi))$. To see this, one may consider a pair of robots i and j, so $G = \{i, j\}$. With E-BEL$^2_{\{i,j\}}$, both robots have:
  - the same intention:
    E-INT$_{\{i,j\}}(\varphi)$;
  - believe they have the same intention:
    the first-order belief E-BEL$_{\{i,j\}}($E-INT$_{\{i,j\}}(\varphi))$; and
  - believe that the other believes this:
    the second-order belief E-BEL$_{\{i,j\}}($E-BEL$_{\{i,j\}}($E-INT$_{\{i,j\}}(\varphi)))$.

Therefore, the robots can reason about the beliefs and intentions of their partner.

In both cases, it is assumed that the robots are incapable of forming coalitions of cardinality ≥ 2. In case 2 the robots will also need to be aware of plan projections of their neighbors, in order to be able to notice when they can help.

Although robots sometimes individually compete for resources, in our application where fast real-time team reaction to dangers is significant, we opt for strictly cooperative robots that use fixed protocols to load up on resources. The clean-up robots do not communicate with robots from other teams and therefore do not need to have any beliefs, intentions and commitments about them.

7.4.3.1 The *UAVs*

The *UAVs* must sometimes work with one another. This requires at least E-BEL^2_G of other *UAVs'* intentions.

The level of intention – within each subteam G_i, the *UAV* $_i$ must make sure that all agents are motivated to do their tasks. Therefore:

- In case 1 we require INT(UAV_i, E-INT$_{G_i}(\varphi)$) with regard to the subteam intention E-INT$_{G_i}(\varphi)$.
- In case 2 we require INT(UAV_i, E-INT$^2_{G_i}(\varphi)$) with regard to the level of subteam intention E-INT$^2_{G_i}(\varphi)$.

The level of belief – within each subteam G_i consisting of an *UAV* $_i$ and $rob_{i_1}, \ldots, rob_{i_{n_i}}$, the *UAV* $_i$ has the highest level of awareness and acts as a coordinator. In order to facilitate this (make plans and reason correctly), it will require one level of belief more than its agents:

- In case 1 we require BEL(UAV_i, E-BEL$_{G_i}$(E-INT$_{G_i}(\varphi)$)) with regard to the subteam's intention E-INT$_{G_i}(\varphi)$.

The same level of awareness is needed with regard to distribution of bilateral commitments within a subteam:

$$\text{BEL}(UAV_i, \text{E-BEL}_{G_i}(\bigwedge_{\alpha \in Cleanup} \bigvee_{i,j \in G} \text{COMM}(i, j, \alpha)))$$

- In case 2 we require BEL(UAV_i, E-BEL$^2_{G_i}$(E-INT$^2_{G_i}(\varphi)$)) with respect to the level of subteam intention E-INT$^2_{G_i}(\varphi)$ as well as:

$$\text{BEL}(UAV_i, \text{E-BEL}^2_{G_i}(\bigwedge_{\alpha \in Cleanup} \bigvee_{i,j \in G} \text{COMM}(i, j, \alpha)))$$

7.4.3.2 The Coordinator

The level of intention– the role of coordinator is to manage the team as a whole (see Figure 7.1), including all subteams and the pilot. Therefore he/she needs to know not only the global plan but also all the subplans. The coordinator has one level of intention more than the *UAVs* it manages and therefore we have INT($coordinator$, INT$^2_G(\varphi)$).

The level of belief–one extra level of belief allows the coordinator introspection and reasoning about the joint effort of *UAVs*. Therefore, since teams are cooperative in a limited way, we have BEL($coordinator$, E-BEL^2_G(E-INT$^2_G(\varphi)$)) with respect to every group intention E-INT$^2_G(\varphi)$. Again, an analogical level of awareness is required with regard to distribution of bilateral commitments.

$$\text{BEL}(coordinator, \text{E-BEL}^2_G(\bigwedge_{\alpha \in Cleanup} \bigvee_{i,j \in G} \text{COMM}(i, j, \alpha))).$$

Commands from the coordinator overrule temporary contracts between teams. It does not only know the plan, but also keeps track of relevant environmental conditions. We assume that even in the safe situation, the robots, the *UAVs* and the pilot are prepared to take action at any moment.

7.4.4   Complexity of the Language Without Collective Attitudes

It seems that in the environmental case study, the language used is richer than propositional modal logic. Fortunately, we can reduce most of the relevant parts to a fixed finite number of propositional atoms (that may be combined and be the subject of attitudes), based on finitely many predicates and constants, as follows:

- a fixed number of relevant environmental states;
- a fixed number of pre-named locations;
- a fixed finite number of agents and teams;
- a fixed finite number of other objects (liquids, solids, catalyst, helicopter);
- a fixed number of relevant thresholds $n_1, n_2, n_3, \varepsilon_1, \varepsilon_2$.

The only possible source of unbounded complexity is the use of continuous intervals and real-valued functions f_1, f_2, f_3, fit appearing in Section 7.2. Recall that the architecture proposed in Section 1.2 allows us to query external entities. These concern data stored in databases and sensed from the environment, which are represented in the lower layer of the system. For example, the functions f_1, f_2, f_3 and fit are part of this lower layer. Therefore, even though the underlying structures are represented by first-order formulas, one extracts only propositional information from them to use in the upper layer of propositional TEAMLOG reasoning. In fact, one can obtain answers *true* or *false* about queries such as:

$$f_3(p(A), t(A), c_1(A), c_2(A)) \in (v_3, n_3]?$$

from the lower layer.

7.5   Conclusion

In this case study we have shown how to implement teamwork within a strictly cooperative, but still heterogenous group of agents in TEAMLOG. The heterogeneity is taken seriously in this application, as advocated in Gold (2005). Natural differences in agents' shares, opportunities and capabilities have been reflected in their awareness about the situation. In fact, the study focused on building beliefs, intentions and commitments of agents involved on an adequate but still minimal level. Even though not all aspects of teamwork have been shown, a bridge between theory and practice of teamwork has been effectively constructed for this exemplary application. Future work will be to embed TEAMLOG into a form of approximate reasoning suitable to model perception in real-world applications. *Similarity-based approximate reasoning* with its intuitive semantics compatible with that of TEAMLOG (Doherty *et al.*, 2007; Dunin-Kęplicz and Szałas, 2007) is a good candidate.

8

Dialogue in Teamwork

Those who know, don't talk.
Those who talk, don't know.

Tao Te Ching (Lao-Tzu, Verse 56)

8.1 Dialogue as a Synthesis of Three Formalisms

Undoubtedly cooperation matters. To make it smart and effective, also communication matters. Its proper realization is very demanding and reflects the art of programing, ensuring an optimal balance between *communication* and *reasoning*. These two complex elements are inevitably present in advanced forms of teamwork. Because we typically deal with problems to be solved collectively by heterogenous agents that are not specifically designed to work together, getting the right team and then controlling its performance is essential, as it was argued already.

In this setting the Contract Net Protocol is often viewed as a simple, but effective and efficient way to distribute tasks over a number of agents aiming for a common goal (Sandholm and Lesser, 1995). It basically makes use of the market mechanism of task demand and supply in order to match tasks with agents willing to perform them. The commercial success of the Contract Net Protocol originates from the use of a fixed protocol with a limited number of steps, which is easy to implement. This market mechanism works well when several agents are willing or even competing to perform tasks that are well described beforehand. However this is rarely the case in multi-agent systems, either because only one agent is capable of performing a given task, and therefore that one should be negotiated with, or because the task cannot be described precisely enough at the very beginning. In such settings, a more refined type of communication cannot be avoided. In particular, advanced forms of teamwork call for subtle, sometimes very complex, but still tractable forms of communication.

In fact, recent models of communication range from rather inflexible communication protocols to more sophisticated constructions based on advanced communication technologies. A good candidate to make conversation between agents flexible is Walton and Krabbe's theory of dialogue (Walton and Krabbe, 1995). Working in the strong tradition

Teamwork in Multi-Agent Systems: A Formal Approach   Barbara Dunin-Kęplicz and Rineke Verbrugge
© 2010 John Wiley & Sons, Ltd

of argumentation theory and informal logic, they distinguish several types of dialogue and give rules for appropriate moves within particular dialogues, without fixing the order of the moves. These moves depend on specific stages of teamwork and most of the time are rather complex.

While dialogues follow the semi-formal theory of Walton and Krabbe, the question how to implement particular moves within these dialogues remains open. It turns out that the well-recognized theory of speech acts of John Austin (see the classic Austin (1975)), later formalized by John Searle and Daniel Vanderveken (see Searle and Vanderveken (1985); Searle (1969)) is a perfect candidate. To put it briefly, speech act theory views communication as complex actions changing the mental states of dialogue participants. Thus, an intuitively appealing method is to realize particular moves by various speech acts, viewed as typical actions. These actions can then be represented in dynamic logic, by characterizing their pre- and post-conditions. Therefore, a basis for a formal system coherent with TEAMLOG is created. Ultimately, the synthesis of the three approaches of *dialogue theory*, *speech act theory* and *dynamic logic* enables us to specify that in given circumstances the dialogue results in a certain outcome. The novelty of the present chapter lies in applying this combination of approaches to a theory of teamwork.

Even though Walton and Krabbe (1995) are not interested in internal attitudes of dialogue participants if these attitudes are not communicated explicitly, modeling the dynamics of teamwork calls for making all aspects of dialogue among computational agents transparent. Therefore, agents' internal attitudes need to be established and then carefully updated and/or revised during teamwork.

In this chapter, we first draw characteristics of particular dialogues. Next, we discuss their role during teamwork. As a reminder, teamwork begins from *potential recognition* when an initiator tries to find out which agents could cooperate on the goal φ and how these can be combined into a team. Secondly, *team formation* is about creating a proper team linked together via C-INT$_G(\varphi)$. Next comes a (possibly collective) *planning phase* resulting in C-COMM$_{G,P}(\varphi)$: the team subdivides the goal, associates subtasks with actions and allocates these to team members. Finally, *team action* is a coordinated execution of individual actions and monitoring the colleagues. One by one we will go through all stages of teamwork.

The chapter is structured in the following manner. Section 8.2 presents characteristics of dialogue types that appear in teamwork. Section 8.3 presents different aspects that play a role in dialogue during teamwork, such as trust, speech acts and Walton and Krabbe's formalization of rigorous persuasion dialogues. The subsequent Sections 8.4, 8.5, 8.6 and 8.7 form the core of this chapter. They present the role of dialogue at each stage of teamwork. The chapter ends with a discussion of recent research on dialogue theory in teamwork and multi-agent systems in general. A significant part of this chapter is based on research with Frank Dignum (Dignum *et al.*, 1999, 2001a,b) and with Alina Strachocka.

8.2   Dialogue Theory and Dialogue Types

Conversations are sequences of messages exchanged between two or more agents. While fixed protocols are too rigid to properly deal with teamwork dynamics, offering complete freedom in communication would be too much for resource-bounded software agents at the current state of communication technology. Therefore a form in between the two extremes, namely dialogue theory, has been *en vogue* recently (Cogan *et al.*, 2005; McBurney *et al.*, 2002; Parsons *et al.* 1998, 2003).

Dialogue theory has been influenced by parallel developments in logic and philosophy in the 1960s and 1970s. Among other researchers, Hintikka (1973), Kambartel (1979), Lorenzen (1961) and Krabbe (2001) developed the idea that semantics of classical and intuitionistic logics could be alternatively formalized in terms of games among two players, instead of the usual Tarski and Kripke semantics. This more dynamical view of semantics has inspired many developments in philosophy, logic and theoretical computer science, for example the invention of dynamic epistemic logics (Baltag *et al.*, 2003, 2008; Benthem, 2001; Ditmarsch *et al.*, 2007).

One of the inventors of dialogue logics, Erik Krabbe, joined forces with Douglas Walton to create a freer version, *dialogue theory*, that is geared more to modeling real-life dialogues than to the semantics of classical or intuitionistic logic. They classified several dialogues: *persuasion, negotiation, inquiry, deliberation, information seeking* and *eristics*, with a special focus on persuasion. As we consider only cooperative teams, eristics, that is verbal fighting between agents, has been left out in the sequel. For each type of dialogue, Walton and Krabbe (1995) formulate *an initial situation, a primary goal* and *a set of rules*. These constitute a *normative model* which is not a record of real dialogues, but represents the ideal way cooperative agents participate in the dialogue in question.

In the course of real-life communication, often a shift from one type of dialogue to another occurs. A special kind of shift, called *embedding*, takes place when the second dialogue is functionally related to the first one and improves its results. For example, persuasion about a certain statement may need an information-seeking phase.

Dialogue theory structures conversations by means of a number of dialogue rules. These rules limit the number of possible responses at each point, while not fixing the sequence of messages. The agents speak in turn, for example asking questions and giving replies and take into account, at each turn, what has occurred previously in the dialogue. The score of the dialogue is kept by each agent as an *attitude store*, to which propositions may be added or retracted during the dialogue in an orderly way. These propositions classically represent informational attitudes like individual beliefs or common beliefs. As a novelty in MAS, they may also represent motivational attitudes like individual goals, individual intentions, collective intentions, social commitments and collective commitments.

Below we shortly explain dialogue theory and briefly describe the speech act theory used to implement the effects of utterances in dialogues between computational agents.

8.2.1 Persuasion

A persuasion dialogue arises from a conflict of opinions. It may be that one agent believes φ while some others either believe a contrary proposition ψ_i (where $\varphi \wedge \psi_i$ is inconsistent) or just have doubt about φ. The goal of a persuasion is to resolve the conflict by verbal means, so as to ensure a stable agreement. In the multi-agent setting, the end result would be a common belief C-BEL$_G(\chi)$, where χ may be the φ or one of the ψ_i, or yet another conclusion resulting from persuasion. Clearly, belief revision takes place here.

Initially, all agents have attitude stores consisting of *theses* and *concessions*. Here, the theses are assertions they are prepared to defend (like φ for the first agent above), while concessions are propositions that are taken for granted for the sake of argument. Walton and Krabbe provide many rules governing effects of updates, revisions and retractions on the attitude stores during persuasion.

In the MAS setting, a *persuasion with respect to motivational attitudes*, not found in Walton and Krabbe, has to be introduced. This new kind of persuasion arises from a conflict of intentions, where one agent intends to achieve φ, while others have a conflicting intention to achieve ψ_i (where φ and ψ_i are inconsistent) or simply lack any positive motivational attitude with respect to φ. The main goal of persuasion with respects to intentions is to resolve this conflict in a way resulting in a stable collective intention.

8.2.2 Negotiation

The initial situation of negotiation is a conflict of interests, together with a need for cooperation. The main goal is to make a deal. Thus, the selling and buying of goods and services often described in the MAS literature is only one of the many contexts where negotiation takes place in multi-agent systems. Negotiation and persuasion are often not distinguished adequately. One has to keep in mind that negotiation is not meant to convince the others of one's viewpoint, as happens in persuasion, but to make a deal leading to a mutually beneficial agreement. There is a wide literature on negotiation in multi-agent systems, covering an area as wide as exchange of services, sale of products and development of treaties among nations (Kraus, 2001; Lin and Kraus, 2008; Sycara, 1990). Formal techniques for negotiation have recently received a lot of attention, from Rosenschein and Zlotkin's *Rules of Encounter* (Rosenschein and Zlotkin, 1994), through information-based negotiation by among others Sierra and Debenham (2007), to game-theoretic approaches by among others Ramchurn *et al.* (2007). We do not go into details here.

In general, Walton and Krabbe (1995) do not allow us to embed negotiation into persuasion, assuming that a proposed statement should be backed by arguments, not offers. When an agent in the course of persuading another agent begins to negotiate, it may be accused of escaping from the burden of proof. Walton and Krabbe call such an illicit embedding of negotiation into persuasion the 'fallacy of bargaining'. On the other hand, persuasion may be fruitfully embedded in negotiation. For example, when setting up the agenda, or in a negotiation about the sale of a house, an embedded persuasion about the market value of similar houses in the neighborhood typically helps clinch the deal.

The rules governing negotiation include severe restrictions on retracting concessions, which are represented mostly as courses of action. In general, when an agent has conceded its willingness to execute some action (for example to sell a product for a certain price) it may not generally retract this concession.

8.2.3 Inquiry

Inquiry starts when some agents are ignorant about the solution to some question or open problem. The main goal is the growth of knowledge, leading to agreement about the conclusive answer of the question. This goal may be attained in many different ways, including an incremental process of argument which builds on established facts in drawing conclusions beyond a reasonable doubt. Both information retrieval and reasoning may be intensively used in this process. The end result of inquiry has a collective flavor and is as strong as C-BEL$_G(\varphi)$ or even C-KNOW$_G(\varphi)$ in some contexts. If one agent reaches an intermediate or final conclusion earlier than others, it may need to persuade them. Therefore, a persuasion dialogue is allowed in inquiry. Conversely, if an open problem appears during persuasion, an inquiry may be embedded to resolve it.

8.2.4 Deliberation

Deliberation as a dialogue is similar to inquiry, but different from both persuasion and negotiation as it starts from an open problem, rather than from a conflict of opinions. Deliberation starts from a need for action performance and is concerned with the future. It aims to reach a decision on how to act in the short term. The kind of reasoning that is central to deliberation and in general to teamwork in multi-agent systems, is *practical reasoning*: goal-directed, knowledge-based reasoning where an agent considers different means of achieving a goal. A typical example of practical reasoning is a means-end analysis linking a particular goal or intention with a, possibly complex, action.

8.2.5 Information Seeking

Information seeking occurs when an agent lacks knowledge on a certain subject or proposition and it seeks this information from others. The end result is a new individual belief BEL(a, φ) of the interested agent a. In contrast to inquiry, the attainment of proof is not essential in information seeking. Apart from collective aspects of inquiry, this distinguishes the two potentially similar dialogues. Information seeking typically occurs in expert consultation, when the questioner has no direct access to information.

8.3 Zooming in on Vital Aspects of Dialogue

In teamwork-related dialogues, both bilateral communication and global announcements take place. As defined in Section 6.2.1, given agents i and j, the action communicate(i, j, ψ) stands for 'agent i communicates to agent j that ψ holds'. Next, given a group G and an agent $i \in G$, the action announce$_G(i, \psi)$ stands for 'agent i announces to group G that ψ holds'.

8.3.1 Trust in Dialogues

Whenever communication between agents appears, the question of trust is inevitably involved. The *trustworthiness* addresses the question 'to what extent does agent j (the receiver) trust agent i (the sender)?'. To make communication and related reasoning more context-sensitive, it is useful to distinguish different gradations of trust. For example, an agent can trust the other entirely (TRUST(j, i) for j trusts i) or with respect to a certain context (for example (TRUST$_\psi(j, i)$ for j trusts i with respect to formula ψ). See Castelfranchi and Falcone (1998), Jøsang *et al.* (2007), Marsh and Dibben (2003) and Ramchurn *et al.* (2004) for interesting discussions about trust in multi-agent systems.

As trust is a rather complex concept, it may be defined in many ways, from different perspectives (see Castelfranchi (2002) and Castelfranchi and Tan (2001) for some relevant work in this area). Though we do not mean to add yet another voice in the ongoing discussion about the role of trust in communication and commonsense reasoning, however, some form of trust has to be adopted in teamwork.

Clearly, it would be too much to assume that agents believe everything communicated to them. Still, for teamwork to succeed it is vital that receivers adopt some information as their own beliefs. For such propositions ψ, the following is justified:

$$succ(\text{communicate}(i, j, \psi)) \rightarrow \text{BEL}(j, \psi)$$

$$succ(\text{announce}_G(i, \psi)) \rightarrow \text{C-BEL}_G(\psi)$$

As long as trust is present in information seeking and inquiry, the speaker's assertions are believed by the hearer (and believed by him/her to be believed by the speaker). Thus, after agent i asserts ψ to agent j in such a context, we have:

$$\text{TRUST}_\psi(j, i) \rightarrow \text{BEL}(j, \psi) \wedge \text{BEL}(j, \text{BEL}(i, \psi))$$

Apparently, for negotiation and persuasion, this need not be the case, as agents do not automatically take on the other's original intentions or beliefs when trying to make a deal or an agreement. To expand on the possible consequences of persuasion on the interlocutors' mental states, Walton and Krabbe's rules for rigorous persuasion are presented and adapted for persuasion with respect to intentions in Section 8.3.3.

In the sequel, we will in some places make the idealizing and simplifying assumption that agents trust one another about everything communicated or announced to them in the course of teamwork. In particular, all agents trust the initiator. Apparently, this assumption may be revised in real multi-agent settings.

8.3.2 Selected Speech Acts

Austin's theory of speech acts, later refined and formalized by Searle and Vanderveken (1985) and Searle (1969), is eminently suitable to account for the influence of a speaker's utterance on the mental state of the hearer. (For an interesting overview of the use of speech act theory in multi-agent systems, see Traum (1999).)

Austin (1975) and Searle (1969) stated that in a speaker's utterance, the agent performs at least the following three kinds of acts:

1. The uttering of words: *utterance acts*.
2. Referring and predicating: *propositional acts*.
3. Stating, questioning, commanding, promising etc.: *illocutionary acts*.

Searle characterized many types of illocutionary acts by four aspects: their propositional content, preparatory conditions, sincerity conditions and essential quality. Our presentation will be restricted to a small set of illocutionary acts that are relevant during potential recognition, team formation and plan formation. These are *assert* ($\text{assert}_{a,i}$), *request* ($\text{request}_{a,i}$), *concede* ($\text{concede}_{a,i}$) and *challenge* ($\text{challenge}_{a,i}$). For request and assert, the four characterizing aspects are defined in Searle (1969); a short reminder comes here.

A *request* has as propositional content a future act α of the hearer. As preparatory condition the hearer must be able to do α and the speaker must believe this; moreover, it should not be obvious to both of them that the hearer will do α anyway. As sincerity condition the speaker must want the hearer to do α. As essential quality a request counts as an attempt to get the hearer to do α.

An *assertion* has as propositional content the stated proposition φ. As preparatory condition the speaker must have reason to believe φ in the current situation. As sincerity condition the speaker must actually believe φ. As essential quality assertion commits the speaker to the truth of φ.

Concessions and *challenges* may be similarly defined. Informally, a concession may be characterized as a hearer's positive reaction to another agent's assertion or request. In the first case, the conceder should believe the other agent's assertion but not necessarily be prepared to defend it. A typical example that we often experience as kids and later as parents is when a mother tries to persuade her toddler to go to sleep by saying 'Look, your teddy bear has already closed its eyes and is falling asleep'. The kid's 'yes' is a concession that needn't be defended.

As a hearer's positive reaction to a request, the concession counts as a promise to fulfil the request; for a full characterization of *promises*, see Searle and Vanderveken (1985) and Searle (1969).

Challenges count as negative reactions to another agent's assertion. As sincerity condition the challenger should not currently believe the propositional content of the assertion, even though it may be persuaded later. Challenges follow the logical structure of the proposition by pointing out a part that is disbelieved.

In addition to the utterance acts, propositional acts and locutionary acts predicated by Austin and Searle, Austin also introduced the notion of the *effects* illocutionary acts have on the actions and attitudes of the hearer. He called such effects *perlocutionary acts*. In Sections 8.4, 8.5 and 8.6, the perlocutionary acts resulting from the speech acts applicable in potential recognition, team formation and plan formation will be defined.

8.3.3 Rigorous Persuasion

Rigorous persuasion is a type of persuasion dialogue that follows formal game rules set up by Walton and Krabbe (1995) in their landmark book *Commitment in Dialogue*. During rigorous persuasion the agents exchange arguments to challenge or support a thesis reflecting their *informational stance*, expressed in terms of beliefs or knowledge. In the course of teamwork, however, it may be essential to persuade another agent to take on a specific intention. This leads to a persuasion towards agents' *motivational stance*.

Rigorous persuasion typically takes place when one wants to persuade someone who is agnostic or even negative about a particular belief or intention. Let us stress, however, that rigorous persuasion should be used sparingly: if there is an easier way to convince an interlocutor, one should go for it.

The following rules adapted from Walton and Krabbe (1995) govern the moves of rigorous persuasion between a proponent (P) and an opponent (O). The two cases of persuading towards beliefs and intentions are distinguished.

8.3.3.1 Rigorous Persuasion with Respect to Beliefs

1. Starting with O the two parties move alternately according to the rules of the game.
2. Each move consists of either a challenge, a question, a statement, a challenge or question accompanied by a statement, or the final remark.
3. The game is highly asymmetrical. All P's statements are assertions, and called *theses*, while all O's statements are called *concessions*. P is doing the questioning, while O does all the challenging.
4. The initial move by O challenges P's initial thesis ψ. It is P's goal to make O concede the thesis. P can do this by questioning O and thus bridging the gap between the initial concessions of O and the thesis or by making an assertion to clinch the argument, if acceptable.
5. Each move for O is to pertain to P's preceding move. If this was a question, then O has to answer it. If it was an assertion, then O has to challenge it.
6. Each party may give up, using the final remark $\mathrm{assert}_{P,O}(quit)$ for the proponent, or $\mathrm{assert}_{O,P}(\mathrm{BEL}(i, \psi))$ for the opponent, where ψ is the belief that P tries to persuade O to take on.

   If O's concessions imply P's thesis, then P is obliged to end the dialogue by the final remark: $\mathrm{assert}_{P,O}(won)$. In our system the following rule is assumed:

   $$[\mathrm{assert}_{P,O}(won)]\mathrm{OBL}(\mathrm{assert}_{O,P}(\mathrm{BEL}(i, \psi)))$$

   Thus, after the proponent asserts his/her success, the opponent is obliged to believe in ψ, and to admit it.
7. All challenges have to follow the logical structure of the thesis. For example, a thesis of the form $A \wedge B$ can be challenged by challenging one of the two conjuncts. For a complete set of rules for the propositional connectives we refer to Walton and Krabbe (1995).

   In the completion stage the outcome of rigorous persuasion is made explicit: either the agents commonly believe in ψ or they know that they differ in opinion.

8.3.3.2 Rigorous Persuasion with Respect to Intentions

1. Starting with O the two parties move alternately according to the rules of the game.
2. Each move consists of either a challenge, a question, a statement, a challenge or question accompanied by a statement, or a final remark.
3. The game is highly asymmetrical. All P's statements are assertions, and called *theses*, while all O's statements are called *concessions*. P is doing the questioning, while O does all the challenging.
4. The initial move by O challenges P's initial thesis. It is P's goal to make O concede the thesis, in this case by taking on the intention to achieve ψ. P can do this by questioning O and thus bridging the gap between the initial concessions of O and the thesis or by making an assertion to clinch the argument, if acceptable.
5. Each move for O is to pertain to P's preceding move. If this move was a question, then O has to answer it. If it was an assertion, then O has to challenge it.
6. Each party may give up, using the final remark $\mathrm{assert}_{P,O}(quit)$ for the proponent, or $\mathrm{assert}_{O,P}(\mathrm{INT}(i, \psi))$, where ψ is the intention that P tries to persuade O to take on.

If O's concessions imply P's thesis, then P is obliged to end the dialogue by the final remark: $\texttt{assert}_{P,O}(won)$.

In our system the following rule is assumed:

$$[\texttt{assert}_{P,O}(won)]\text{OBL}(\texttt{assert}_{O,P}(\text{INT}(i, \psi)))$$

which means that after the proponent asserts his/her success, the opponent is obliged to state that he has been persuaded and takes on the intention to achieve ψ.

7. All challenges have to follow the logical structure of the formulas in question.

In the next four sections, we will concentrate on different dialogues involved in the realization of particular stages of teamwork.

8.4   Information Seeking During Potential Recognition

Potential recognition is about finding a set of agents that are prepared to cooperate towards a common goal. These agents are grouped into a sequence of potential teams with whom further discussion will follow during *team formation*. As a reminder, we assume that there is one initiator between them. The first task of the *initiator a* is to form a partial (abstract) plan leading to the goal. On the basis of the (type of) recognized subgoals it will determine which agents might be most suited to form the team. To determine this match, the initiator tries to find out the properties of agents, being interested in four aspects, namely their *abilities*, *opportunities*, *willingness* to participate in team formation and their *type*. Ultimately, the initiator has to form beliefs about the abilities, opportunities, the willingness and the distribution of types of the individual agents in order to derive PotCoop(a, φ). Here is a reminder of the formula (see Chapter 5 for an extensive discussion):

$$\text{PotCoop}(\varphi, a) \leftrightarrow \text{GOAL}(a, \varphi) \wedge$$

$$\exists G \subseteq T \ (\text{BEL}(a, c\text{-}can_G(\varphi) \wedge \bigwedge_{i \in G} willing(i, \varphi) \wedge$$

$$propertypedistr(G, \varphi))) \wedge$$

$$(\neg can(a, \varphi) \vee \neg\text{GOAL}(a, done(a, \texttt{stit}(\varphi))))$$

The initiator can gather the necessary information by asking every agent about its properties and the agent responding with the requested information. Formally this can be expressed by the request scheme below. One can express the 'if ψ **then** α **else** β' construction in dynamic logic, by:

$$(\texttt{confirm}(\psi); \alpha) \cup (\texttt{confirm}(\neg\psi); \beta)$$

In the sequel, we will use the more legible abbreviations with **if** ... **then** ... **else** ...:

$$\texttt{request}_{a,i}(\textbf{if } \psi \textbf{ then } \texttt{assert}_{i,a}(\psi) \textbf{ else } \texttt{assert}_{i,a}(\neg\psi))$$

where $\texttt{request}_{a,i}(\alpha)$ stands for agent a requesting agent i to perform the action α. Thus, in the formal request above, a requests i to assert ψ if ψ is the case and to assert $\neg\psi$ if not. During potential recognition, ψ may stand for any formula of the forms:

- $able(i, \psi_i)$;
- $opp(i, \psi_i)$;
- $willing(i, \varphi)$;
- $type(i, blindly\text{-}committed)$ (similarly for other agent types).

After this request i has four options:

1. It can simply ignore a and not answer at all.
2. It can state that it is not willing to divulge this information:
   $\text{assert}_{i,a}(\neg(\textbf{if } \psi \textbf{ then } \text{assert}_{i,a}(\psi) \textbf{ else } \text{assert}_{i,a}(\neg\psi)))$.
3. It can state that it does not have enough information:
   $\text{assert}_{i,a}(\neg(\text{BEL}(i, \psi) \wedge \neg\text{BEL}(i, \neg\psi)))$.
4. It can either assert that ψ is the case or that it is not.

Of course in case 2, agent a can already derive that i is not willing to achieve φ as part of a team; only in case 4 will a have a resulting belief about ψ.

The formula below represents the result of a sequence of utterances, under the assumption that there is trust with respect to the relevant proposition. The formula is based on a dynamic logic formula of the form $[\alpha_1][\alpha_2]\psi$, meaning that if α_1 is performed then always a situation arises such that if α_2 is performed then in the resulting state ψ will always hold (see Section 6.3.1 for a short reminder of dynamic logic and its use in TEAMLOG^{dyn}).

Therefore, the formula below shows the update of the initiator's mental state after a positive answer, in words: after initiator a requests agent i to answer by asserting whether or not ψ, and after i's positive reply asserting that indeed ψ, then if the initiator trusts i with respect to ψ, the initiator will adopt the belief in ψ and will also believe that i believes ψ:

$$[\text{request}_{a,i}(\textbf{if } \psi \textbf{ then } \text{assert}_{i,a}(\psi) \textbf{ else } \text{assert}_{i,a}(\neg\psi))]$$

$$[\text{assert}_{i,a}(\psi)](\text{TRUST}_\psi(a, i) \rightarrow \text{BEL}(a, \psi) \wedge \text{BEL}(a, \text{BEL}(i, \psi)))$$

After a negative answer, on the other hand, the initiator's beliefs are updated negatively as well: after initiator a requests agent i to answer by asserting whether or not ψ, and after i's negative reply asserting that $\neg\psi$, then if the initiator trusts i with respect to ψ, the initiator will adopt the belief in $\neg\psi$ and will also believe that i believes $\neg\psi$:

$$[\text{request}_{a,i}(\textbf{if } \psi \textbf{ then } \text{assert}_{i,a}(\psi) \textbf{ else } \text{assert}_{i,a}(\neg\psi))]$$

$$[\text{assert}_{i,a}(\neg\psi)](\text{TRUST}_\psi(a, i) \rightarrow \text{BEL}(a, \neg\psi) \wedge \text{BEL}(a, \text{BEL}(i, \neg\psi)))$$

The role of Trust in the Information Seeking Dialogue

Awareness of trust makes a difference in the consequences for the agents' mental states. Thus, if i believes that the initiator trusts it, the part

$$\text{TRUST}_\psi(a, i) \rightarrow \text{BEL}(a, \psi) \wedge \text{BEL}(a, \text{BEL}(i, \psi))$$

may be adapted to derive a different, higher-order, conclusion. Thus, an update of mental states is performed:

[request$_{a,i}$ (**if** ψ **then** assert$_{i,a}(\psi)$ **else** assert$_{i,a}(\neg\psi))$]

[assert$_{i,a}(\psi)$](BEL$(i,$ TRUST$_\psi(a,i)) \rightarrow$ BEL$(i,$ BEL$(a, \psi) \wedge$ BEL$(a,$ BEL$(i, \psi))))$

If the initiator's trust of i is commonly believed by both agents, a much stronger conclusion may be derived resulting in mutual awareness about the mental state of both sides involved in a dialogue:

[request$_{a,i}$ (**if** ψ **then** assert$_{i,a}(\psi)$ **else** assert$_{i,a}(\neg\psi))$]

[assert$_{i,a}(\psi)$](C-BEL$_{i,a}$(TRUST$_\psi(a,i)) \rightarrow$ C-BEL$_{\{i,a\}}(\psi))$

Tactics of Information Seeking

During information seeking about the ingredients of PotCoop(a, φ) the schema of all necessary questions may be rather complex if one conforms to efficiency and complexity standards in Computer Science. For example, in order to recognize the ability to achieve a specific subgoal φ_i, agent a should repeat this question with respect to all distinguished subgoals to every agent. Clearly, such a solution is not acceptable. It is more effective to ask each agent to divulge all its abilities regarding the entire set of goals, as an exemplary solution from the wide spectrum of possibilities. Because these strategic considerations are not related directly to the theory of dialogue, they will be left out.

A next strategic point deals with case 1 above: to avoid the initiator waiting indefinitely for an answer, we incorporate an implicit deadline for reaction for any speech act. After this deadline, the silent agent will not be considered as a potential team member anymore. In fact, this seems to be a very effective solution in agents' communication, even if not our favorite in everyday life. The logical modeling of these types of deadlines is described in Dignum and Kuiper (1998) and will not be pursued here.

End Result of Potential Recognition

Finally, the successful result of potential recognition is that agent a is positive about forming a team aiming to realize φ:

$$\text{BEL}(a, c\text{-}can_G(\varphi) \wedge \bigwedge_{i \in G} willing(i, \varphi) \wedge propertypedistr(G, \varphi))$$

where the initiator holds these positive beliefs for groups $G \subseteq T$. This information should be divulged to all agents by the initiator broadcasting the end result of potential recognition. The effects on the individual and collective mental states of the agents involved may be given in a way similar to the two-agent communications presented above.

As a reminder, we suppose that the communication medium is commonly believed to be perfect and that the initiator is commonly believed to be perfectly trustworthy.

Then we have for the relevant potential groups G:

$$[\text{announce}_{a,G}(c\text{-}can_G(\varphi) \wedge \bigwedge_{i \in G} willing(i, \varphi) \wedge propertypedistr(G, \varphi))]$$

$$\text{C-BEL}_G(c\text{-}can_G(\varphi) \wedge \bigwedge_{i \in G} willing(i, \varphi) \wedge propertypedistr(G, \varphi))$$

At any rate, if all agents in G trust a with respect to

$$(c\text{-}can_G(\varphi) \wedge \bigwedge_{i \in G} willing(i, \varphi) \wedge propertypedistr(G, \varphi)),$$

all of them will believe in this after an announcement and assuming that this trust is a common belief, the content of the announcement is commonly believed as well. Finally the success of potential recognition is commonly believed in the relevant potential groups.

8.5 Persuasion During Team Formation

During potential recognition, individual properties of agents that are essential for cooperation (for example, the exchange of services) were considered. Then, team formation transforms a loosely coupled group into a strictly cooperative team. As a reminder, during team formation the initiator attempts to bring it about that in some group G agents have a *collective intention* (see Section 3.5) to achieve φ. The input of this stage is an initiator a, a goal φ and collection of potential groups. The output of team formation is a selected group G, together with a collective intention C-INT$_G(\varphi)$.

Note that this concept of teamwork requires agents that have a type of 'social conscience'. We do not view a set of agents as a team if they cooperate by just achieving their own predefined part of a common goal. If agents are part of a team, they should be interested in the performance of the other team members and willing to adjust their task to the needs of others. In fact, such a subtle adjustment calls for rather refined dialogues. At the beginning, the initiator keeps a collection of groups in mind. All members of these potential teams have expressed their willingness to participate towards the common goal but do not necessarily have their relevant individual intentions yet. In this situation, the initiator needs to persuade them to take on these intentions and to act together as a team.

8.5.1 Creating Collective Intention

The main type of dialogue during team formation is *persuasion with respect to motivational attitudes*. This arises from a potential conflict of intentions between interested agents or simply from a lack of any positive motivational attitude with respect to φ. The persuasion is mostly one-sided so that in the end the initiator a has persuaded all agents to adopt the intention to work together.

In contrast to persuasion with respect to beliefs, bargaining may be appropriate within a persuasion with respect to goals or intentions. For example, during team formation potential team members may be reasonably persuaded using an embedded negotiation about return favors from agent a.

8.5.1.1 Goal of the Persuasion Dialogue

The goal of the persuasion dialogue is to establish a collective intention towards φ (C-INT$_G(\varphi)$). We recall here the axioms for mutual and collective intentions (see also Section 3.5):

M1 E-INT$_G(\varphi) \leftrightarrow \bigwedge_{i \in G}$ INT(i, φ)
M2 M-INT$_G(\varphi) \leftrightarrow$ E-INT$_G(\varphi \wedge$ M-INT$_G(\varphi))$
M3 C-INT$_G(\varphi) \leftrightarrow$ M-INT$_G(\varphi) \wedge$ C-BEL$_G($M-INT$_G(\varphi))$

Axiom **M2** makes evident that the initiator needs to persuade all potential team members, firstly, to accept the main goal as individual intention and secondly, to accept the intention towards a mutual intention to contribute to this goal, in order to foster cooperation from the start. It suffices if the initiator persuades all potential team members to take on an individual intention towards φ and the intention that there be a mutual intention among that team.

Formally, for all $i \in G$, the initiator seeks to establish INT$(i, \varphi \wedge$ M-INT$_G(\varphi))$. For this implies by axiom **M1** that E-INT$_G(\varphi \wedge$ M-INT$_G(\varphi))$, which in turn implies by axiom **M2** that M-INT$_G(\varphi)$. When all the individual motivational attitudes are established within G, the initiator broadcasts the fact M-INT$_G(\varphi)$, by which the necessary common belief C-BEL$_G($M-INT$_G(\varphi))$ is in place, ensuring the collective intention.

This will be achieved during a persuasion dialogue, which according to Walton and Krabbe (1995) consists of three main stages: *information exchange*, *rigorous persuasion* and *completion*. In our case the information exchange already started during potential recognition. The team formation succeeds when for one potential team all the persuasion dialogues have been concluded successfully.

8.5.2 Agents Persuading One Another to Join the Team

Intentions are formed on the basis of beliefs and previously formed high-level intentions by a number of generic rules (see Dignum and Conte, 1997). For example, the built-in intention can be to obey the law or to avoid punishment. The (instrumental) belief is that driving slower than the speed limit is instrumental for obeying the law and is the *preferred* way to do so. Together with the intention generation rule, the new intention of driving slower than the speed limit is derived.

The general intention generation rule may be represented as follows:

$$\text{INT}(i, \psi) \wedge \text{BEL}(i, \text{INSTR}(i, \chi, \psi)) \wedge \text{PREFER}(i, \chi, \psi) \rightarrow \text{INT}(i, \chi) \quad (8.1)$$

It states that if an agent i intends to achieve ψ, and it believes that χ is instrumental in achieving ψ, and χ is its preferred way of achieving ψ, then it will have the intention to achieve χ. The statement 'χ is instrumental in achieving ψ' means that achieving χ gets the agent 'closer' to ψ in some abstract sense. We do not refine this relation any further, but leave it as primitive.

The PREFER relation is based on an agent's individual beliefs about the utility ordering between its goals, collected here into a finite set H. We abstract from the specific way in which the agent may compute the relative utilities (see the literature about qualitative decision theory (Boutilier, 1994)). An alternative qualitative account is given in van Benthem and Liu (2007).

$$\text{PREFER}(i, \chi, \psi) \equiv$$

$$\bigwedge_{\xi \in H} (\text{BEL}(i, \text{INSTR}(i, \xi, \psi)) \rightarrow \text{BEL}(i, util(i, \chi) \geq util(i, \xi))) \qquad (8.2)$$

Thus, χ is the preferred way for agent i to achieve ψ, if among all goals that are instrumental for achieving ψ, goal χ has the highest utility for i.

8.5.2.1 Information Exchange During Persuasion

During information exchange the agents make clear their initial stand toward the possibility of teamwork. These issues are expressed partly in the form of intentions and beliefs. Other supporting or related beliefs might also be exchanged already. In order not to waste time and energy, a full-fledged persuasion dialogue only needs to take place in case a real conflict arises.

In each persuasion, there are two parties or roles; the proponent (P) and the opponent (O). In the sequel the proponent P is played by the initiator a and the opponent O by the agent i it interacts with. The stands of opponent O are seen as its initial *concessions*. Concessions are beliefs and intentions that an agent takes on for the sake of argument but need not be prepared to defend. Naturally, the agents will also have other private attitudes that may appear later in the course of the dialogue. The stand of the initiator (P) is the goal ψ it is trying to let O take on and which it is prepared to defend during dialogue. The initial conflict description consists of the set of O's initial concessions and P's intention ψ. (For the rules for rigorous persuasion with respect to intentions, see Section 8.3.3.)

In step 6 of the rigorous persuasion game, the successful result for the initiator a would be that his interlocutor i gives up by making the following assertion:

$$\texttt{assert}_{i,a}(\text{INT}(i, \varphi \wedge \text{M-INT}_G(\varphi)))$$

This means that i accepts its role in the team by asserting that it takes on the intention to achieve φ, not alone but together with team G: i also takes on the intention that there be a mutual intention in the team.

8.5.3 Speech Acts and their Consequences During Persuasion

In contrast to different settings, for example Walton and Krabbe (1995), during teamwork we need to monitor agents' informational and motivational attitudes during persuasion. In the course of dialogue we are concerned with assertions and challenges with respect to beliefs, and concessions and requests with respect to both beliefs and intentions.

8.5.3.1   Consequences of Assertions

As for assertions, after a speech act of the form $\texttt{assert}_{a,i}(B)$, standing for 'agent a asserts statement B to agent i', agent i naturally believes that the initiator believes that B:

$$[\texttt{assert}_{a,i}(B)]\mathrm{BEL}(i, \mathrm{BEL}(a, B)) \tag{8.3}$$

Let us assume that i has two rules for answering an assertion B. If i does not have a belief that is inconsistent with B then i will concede, so B's consistency with the agent's beliefs has a role similar to that of justifications in default logic (Antoniou, 1997; Łukaszewicz, 1990; Reiter, 1980):

$$\neg\mathrm{BEL}(i, \neg B) \rightarrow do(i, \texttt{concede}_{i,a}(B)) \tag{8.4}$$

If, on the other hand, i believes in the contrary, it will challenge the assertion:

$$\mathrm{BEL}(i, \neg B) \rightarrow do(i, \texttt{challenge}_{i,a}(B)) \tag{8.5}$$

where the operator $\mathrm{DO}(i, \alpha)$ indicates that α is the next action performed by i.

8.5.3.2   Consequences of Concessions

The $\texttt{concede}$ action with respect to beliefs is basically an assertion plus a possible mental update of the agent. In effect, the agent does not only assert the proposition but actually believes it, even if this was not the case beforehand. Suppose that i did not have a contrary belief, then i concedes B by the speech act $\texttt{concede}_{i,a}(B)$ with the effect similar to $\texttt{assert}$, except that a can only assume that i believes B during the dialogue and might retract it afterwards.

$$[\texttt{concede}_{i,a}(B)]\mathrm{BEL}(a, \mathrm{BEL}(i, B)) \tag{8.6}$$

8.5.3.3   Consequences of Challenges

The $\texttt{challenge}$ with respect to beliefs is a combination of a denial of the proposition (assertion of a belief of the negated proposition) and a request to prove the proposition. The exact form of the challenge depends on the logical form of the proposition in question (Walton and Krabbe, 1995). Thus, the complete effects of this speech act are quite complex. An exemplary challenge will be described in Section 8.5.5.

8.5.3.4   Consequences of Persuasion with Respect to Intentions

In the case of persuasion with respect to intentions, the situation is different. For example, initiator a requests i to take on an intention ψ by the following speech act:

$$\texttt{request}_{a,i}(\texttt{concede}_{i,a}(\mathrm{INT}(i, \psi)))$$

Similar to the case of assertions, i has two rules for answering such a request. If i does not have a contradicting intention to achieve $\neg\psi$, then i will concede:

$$\neg\mathrm{INT}(i, \neg\psi) \rightarrow do(i, \texttt{concede}_{i,a}(\mathrm{INT}(i, \psi))) \tag{8.7}$$

Here, i concedes by the speech act $\texttt{concede}_{i,a}(\text{INT}(i, \psi))$ resulting in an effect on the initiator's mental state:

$$[\texttt{concede}_{i,a}(\text{INT}(i, \psi))]\text{BEL}(a, \text{INT}(i, \psi)) \tag{8.8}$$

In words, this means that always when i concedes to a that it intends ψ, then a believes that i indeed intends ψ.

If, in contrast, i does have a contrary intention to achieve $\neg\psi$, it will assert that it indeed intends $\neg\psi$:

$$\text{INT}(i, \neg\psi) \rightarrow do(i, \texttt{assert}_{i,a}(\text{INT}(i, \neg\psi))) \tag{8.9}$$

The result of the assertion on the initiator's mental state is captured by:

$$[\texttt{assert}_{i,a}(\text{INT}(i, \neg\psi))]\text{BEL}(a, \text{INT}(i, \neg\psi)) \tag{8.10}$$

This means that always when i asserts to a that it intends $\neg\psi$, then a believes that i indeed intends $\neg\psi$.

8.5.4 Announcing the Success of Team Formation

When all the individual motivational attitudes are established within the team G, meaning that a has persuaded all $i \in G$ to take on $\text{INT}(i, \varphi \wedge \text{M-INT}_G(\varphi))$, the initiator broadcasts the fact:

$$\text{E-INT}_G(\varphi) \wedge \text{E-INT}_G(\text{M-INT}_G(\varphi))$$

The result of this broadcast depends on the degree of trust in the initiator present among the agents in G. Thus, we have (leaving out the long formula as subscript of TRUST):

$$[\texttt{announce}_{a,G}(\text{E-INT}_G(\varphi) \wedge \text{E-INT}_G(\text{M-INT}_G(\varphi)))]$$

$$(\bigwedge\nolimits_{i \in G} \text{TRUST}(i, a) \rightarrow \bigwedge\nolimits_{i \in G} \text{BEL}(i, \text{E-INT}_G(\varphi) \wedge \text{E-INT}_G(\text{M-INT}_G(\varphi)))$$

Note that when everybody trusts the initiator, but there is no common belief about this, a collective intention is not quite achieved. Positively, when the initiator is commonly believed to be trustworthy, we have in general:

$$[\texttt{announce}_{a,G}(\psi))]$$

$$(\text{C-BEL}_G(\text{TRUST}(i, a, \psi)) \rightarrow \text{C-BEL}_G(\psi))$$

Therefore, for $\psi = \text{E-INT}_G(\varphi)) \wedge \text{E-INT}_G(\text{M-INT}_G(\varphi))$, we have:

$$[\texttt{announce}_{a,G}(\text{E-INT}_G(\varphi)) \wedge \text{E-INT}_G(\text{M-INT}_G(\varphi)))]$$

$$(\text{C-BEL}_G(\text{TRUST}(i, a)) \rightarrow \text{C-BEL}_G(\text{E-INT}_G(\varphi) \wedge \text{E-INT}_G(\text{M-INT}_G(\varphi))))$$

In words, after the initiator announces to group G that everyone intends to achieve φ and everyone intends to achieve it together by $\text{M-INT}_G(\varphi)$, then, as long as trust in the initiator is commonly believed, the result will be a common belief in the group that everyone intends the needed ingredients. In this way, the necessary common beliefs are established and, by the reasoning of Section 8.5.1, the *collective intention* C-INT$_G(\varphi)$ is in place. The initiator has succeeded in creating a team, ready to start planning how to achieve the goal.

8.5.5   Team Formation Through the Magnifying Glass

Let us consider an exemplary case of team formation for achieving the following goal φ:

> To arrange a trip of three weeks to Australia for a certain famous family; the trip should satisfy specific constraints on costs, times, places and activities.

The initiative for teamwork is taken by travel agent a, who cannot arrange the whole trip on its own. The trip will be extensively publicized, so it has to be a success, even if circumstances change. Hence, it does not simply ask airline companies, hotels and organizers of activities to deliver a fixed combination of services. Instead, it believes that a more flexible type of true teamwork gives the best chances of a successful trip.

During team formation, the initiator tries to persuade the other agents i in the potential team to take on the intention to achieve the overall goal of organizing the journey (INT(i, φ)), but also with respect to doing this as a team (INT(i, M-INT$_G(\varphi)$)). To this end, the initiator exploits the theory of intention formation.

The mechanism sketched in Section 8.5.3 can be used during persuasion. The initiator tries to get the other agents in G to concede to higher-level intentions, instrumental beliefs and preferences that together with Rule (8.1) imply the intention to achieve the overall goal φ (as proposed in Dignum and Weigand, 1995). To be more concrete, the higher-level intention ψ could stand for 'earn good money'. Here follows an example move of the initiator:

$$\text{assert}_{a,i}(\bigwedge_{j \in G} (\text{INT}(j, \psi) \rightarrow \text{INSTR}(j, \varphi, \psi)))$$

Thus, the initiator states that if an agent has the higher-level intention to earn good money, then φ is instrumental for achieving this. After this speech act agent i believes that the initiator believes what it asserts, according to general Rule (8.3), therefore:

$$[\text{assert}_{a,i}(\bigwedge_{j \in G} (\text{INT}(j, \psi) \rightarrow \text{INSTR}(j, \varphi, \psi)))]$$

$$\text{BEL}(i, \text{BEL}(a, \bigwedge_{j \in G} (\text{INT}(j, \psi) \rightarrow \text{INSTR}(j, \varphi, \psi))))$$

According to the general discussion about consequences of assertions in Section 8.5.3, there are two possibilities for i's answer. Let us assume that the positive case holds, that is i does not have a contrary belief and so it concedes by Rule (8.4):

$$\text{concede}_{i,a}(\bigwedge_{j \in G} (\text{INT}(j, \psi) \rightarrow \text{INSTR}(j, \varphi, \psi)))$$

The effect of this speech act on the initiator follows by general Rule (8.6):

$$[\text{concede}_{i,a}(\bigwedge_{j \in G} (\text{INT}(j, \psi) \to \text{INSTR}(j, \varphi, \psi)))]$$

$$\text{BEL}(a, \text{BEL}(i, \bigwedge_{j \in G}(\text{INT}(j, \psi) \to \text{INSTR}(j, \varphi, \psi))))$$

Now the formula is believed by both a and i. Thus, the initiator's next aim in the persuasion will be to get i to intend ψ (earn good money) by the question:

$$\text{request}_{a,i}(\text{concede}_{i,a}(\text{INT}(i, \psi)))$$

By the general Rules (8.7) and (8.9), i is obliged to either concede to take on the intention ψ (if this is consistent with its other intentions) or to assert that it intends its negation. After i's response, the initiator believes i's answer. Note that in the second case, it may be useful for a to embed a negotiation dialogue in the persuasion, in order to get i to revise some of its previous intentions.

When the initiator has persuaded agent i to take on the high-level intention ψ and to believe the instrumentality of φ with respect to ψ, it can go on to persuade the other that φ is its preferred way of achieving ψ ($\text{PREFER}(i, \varphi, \psi)$) by a speech act constructed according to Definition 8.2:

$$\text{assert}_{a,i}(\bigwedge_{\xi \in H}(\text{BEL}(i, \text{INSTR}(i, \xi, \psi)) \to \text{BEL}(i, util(i, \varphi) \geq util(i, \xi)))) \quad (8.11)$$

Here, H is a pre-given set of propositions representing instrumental goals. Note that the 'first-order' part of this formula, $util(i, \varphi) \geq util(i, \xi)$, refers to the lower layer of the architecture described in Section 1.12, where comparisons between real numbers can be done, and a yes–no output is transferred to the upper layer of TEAMLOG reasoning. Here the first-order part is included in the formula to make the presentation clearer but note that overall, we still work in a propositional theory.

To make the example more interesting, suppose that i does not yet prefer φ as a means to earn good money. Instead it believes that χ, arranging some less complex holidays for another family, has a higher utility than φ. Thus i does not concede to the initiator's speech act but instead counters with a challenge. According to the logical structure of the assertion in Rule (8.11), this challenge is a complex speech act consisting of three consecutive steps. First, i asserts the negation of a's assertion, namely:

$$\neg(\bigwedge_{\xi \in H}(\text{BEL}(i, \text{INSTR}(i, \xi, \psi)) \to \text{BEL}(i, util(i, \varphi) \geq util(i, \xi))))$$

which is a conjunction of implications. Then it concedes to the antecedent:

$$\text{BEL}(i, \text{INSTR}(i, \chi, \psi))$$

of the implication for the specific goal $\chi \in H$. And finally it requests a to present a proof that φ has a better utility for i than χ. Summing up:

$\text{challenge}_{i,a}$

$$(\bigwedge_{\xi \in H} (\text{BEL}(i, \text{INSTR}(i, \xi, \psi)) \to \text{BEL}(i, util(i, \varphi) \geq util(i, \xi)))) \equiv$$

$$\text{assert}_{i,a}(\neg(\bigwedge_{\xi \in H}(\text{BEL}(i, \text{INSTR}(i, \xi, \psi)) \to \text{BEL}(i, util(i, \varphi) \geq util(i, \xi)))));$$

$$\text{concede}_{i,a}(\text{BEL}(i, \text{INSTR}(i, \chi, \psi)));$$

$$\text{request}_{i,a}(\text{assert}_{a,i}(\text{PROOF}(util(i, \varphi) \geq util(i, \chi))))$$

As a reply, a could prove that the utility of φ is in fact higher than that of χ, because it generates a lot of good publicity, which will be profitable for i in future – something of which i was not yet aware. Let us suppose that i is persuaded by the argument and indeed concedes to its new preference by the speech act:

$$\text{concede}_{i,a}(\text{PREFER}(i, \varphi, \psi))$$

All these concessions, together with the general intention formation Rule (8.1) and the fact that agents are correct about their intentions, then lead to $\text{INT}(i, \varphi)$. For intentions about cooperation with other potential team members, the process to persuade the agent to take on $\text{INT}(i, \text{M-INT}_G(\varphi))$ is analogous.

8.6 Deliberation During Planning

In the AI and MAS literature many methods of planning have been developed. The essential characteristics of BDI agents developed in this book is that they are capable of *planning from first principles*, including the phases of *task division*, *means-end analysis* and *action allocation*, discussed in detail in Chapter 5. As a reminder, the input of plan formation is a team G together with C-INT$_G(\varphi)$. The successful outcome is formula C-COMM$_{G,P}(\varphi)$ (a collective commitment of the group G based on the social plan P).

During planning from first principles, *deliberation* is typically the essential dialogue, including various embeddings, as discussed in the sequel.

8.6.1 Stages of Deliberation: Who Says What and with Which Effect?

The aim of a deliberation dialogue is to make a common decision of what to do in the near future. Therefore this dialogue needs both a formal opening specifying its subject as well as a formal closure confirming the decision made. The definition below of the deliberation stages, speech acts and their semantics benefit from the formal model proposed by McBurney *et al.*, (2007) and from the master's thesis on deliberation by Alina Strachocka of Warsaw University.

There are two parties of the dialogue: the initiator a, and the rest of the team G. From a game-theoretic perspective, deliberations are asymmetric games because a has more moves at its disposal than the other team members. Deliberation consist of four phases:

Opening,   when the subject of the dialogue is presented and the dialogue is opened;
Voting,   when the opinions of the dialogue participants are collected;
Confirming,   when the proposed decision is announced and possibly counter-arguments
                  are raised;
Closure,   when the final decision is announced.

In each phase different speech acts from assert, concede, challenge and request occur, as characterized in the previous sections. If the problem in question can be described as a formula, for example $\psi(x)$, then the aim of the deliberation dialogue among G on the subject '$\psi(x)$' is to find the best t satisfying $\psi(x)$ from a finite given set of candidates T_ψ and to achieve C-BEL$_G(\psi(t))$.

The structure of deliberation is imposed by a, while the rest of the agents in G react accordingly. Initiator a leads deliberation based on the following rules for the four phases.

8.6.1.1 Opening

The deliberation dialogue on subject ψ is opened by a's request to all other $i \in G$:

$$\text{request}_{a,i}(\textbf{if } \bigvee_{t \in T_\psi} \psi(t) \textbf{ then } \text{assert}_{i,a}(\psi(t)) \textbf{ else } \text{assert}_{i,a}(\neg \bigvee_{t \in T_\psi} \psi(t)))$$

According to the rules about requests in Section 8.4, the other agents in G have four options of reacting. Agent a waits for a certain amount of time before collecting all the answers from G. If no agent provides an answer in time, deliberation fails.

8.6.1.2 Voting

The second step consists in the initiator announcing all answers collected from the agents during the first step or a pre-selected subset of them. Let $T_{\psi,a} \subseteq T_\psi$ denote this finite set of selected candidate terms of a with respect to ψ. Then a divulges the candidate set to each colleague $i \in G$ by:

$$\text{assert}_{a,i}(\bigwedge_{t \in T_{\psi,a}} \bigvee_{i \in G} \text{BEL}(i, \psi(t)))$$

Subsequently, a opens the voting by requests to all $i \in G$:

$$\text{request}_{a,i}(\bigwedge_{x,y \in T_{\psi,a}} (\textbf{if } \psi(x) \wedge \text{PREFER}(i, x, y) \textbf{ then}$$

$$\text{assert}_{i,a}(\text{PREFER}(i, x, y))))$$

Here, PREFER(i, x, y) stands for 'i prefers option x above option y'. Just like in the previous step, the agents have four options for their reactions. In case no one answers, the scenario leads back to step one, the opening. The communication during voting could have caused some belief revisions among the agents and therefore the return to phase one is justified. Otherwise, if some answers are received, a counts the votes, possibly applying different evaluation functions, which may, for example, be weighted with respect to trust towards particular agents.

8.6.1.3   Confirming

In step 3, a announces the winning proposal w and calls the agents that have something against it to start a persuasion dialogue, using a request to all other $i \in G$:

$$\mathtt{request}_{a,i} \ (\textbf{if } \text{BEL}(i, \neg\psi(w)) \vee \bigvee_{t \in T_\psi} (\text{PREFER}(i, t, w)) \ \textbf{then}$$

$$\mathtt{assert}_{i,a}(\neg\psi(w) \vee \text{PREFER}(i, t, w)))$$

During this phase, if no agent steps out, the scenario moves to the final stage, the closure. If, on the other hand, there is an agent j who thinks that the chosen option is not the best one, then j has to announce this. The reaction of j may lead to a `challenge` by a to provide a proof, by which the dialogue switches to persuasion: j must either convince a of its own preferred candidate t or it must convince a that $\psi(w)$ does not hold. If j is successful, a takes on j's preference and announces j's thesis to all other $i \in G$:

$$\mathtt{assert}_{a,i}(\neg\psi(w) \vee \text{PREFER}(a, t, w))$$

In this situation, the remaining agents may concede:

$$\mathtt{concede}_{i,a}(\neg\psi(w) \vee \text{PREFER}(i, t, w))$$

or challenge the thesis:

$$\mathtt{challenge}_{i,a}(\neg\psi(w) \vee \text{PREFER}(i, t, w))$$

If anyone chooses to challenge, a must enter into a persuasion dialogue with the challenger. Finally, the scenario returns to the preceding voting step and the agents have an opportunity to express their changed opinions.

8.6.1.4   Closure

In step 4, a announces the final choice, z, by the complex speech act $\mathtt{announce}_{a,G}(\psi(z))$, consisting of multiple repeated assertions to all other $i \in G$:

$$\mathtt{assert}_{a,i}(\psi(z) \wedge \mathit{done}(\mathtt{announce}_{a,G}(\psi(z))))$$

Its consequences depend on the consequences of the particular component speech acts, which in turn depend on the level of trust towards a.

The assertion of $\mathit{done}(\mathtt{announce}_{a,G}(\psi(z)))$ is essential for achieving the common belief $\text{C-BEL}_G(\psi(z))$, similar to the protocol discussed in Section 2.7.3. Apart from informing about ψ, the initiator delivers a message that it just uttered an announcement of $\psi(z)$ to the whole group G. After these assertions, the usual consequences for other agents' belief states ensure:

$$\text{BEL}(i, \psi(z)) \wedge \mathit{done}(\mathtt{announce}_{a,G}(\psi(z)))$$

and:

$$\text{BEL}(i, \text{BEL}(a, \psi(z) \wedge done(\text{announce}_{a,G}(\psi(z)))))$$

In other words, agent i believes that $\psi(z)$ was announced to the group (and answers with a `concede` about both $\psi(z)$ and the announcement), but may not have information whether the others trust a, nor therefore whether they adopt the belief $\psi(z)$.

In the case group G commonly believes that a is trustworthy, however, the agents know what consequences the announcements will bring about for the other agents:

$$[\text{announce}_{a,G}(\psi(z)) \wedge done(\text{announce}_{a,G}(\psi(z)))]$$

$$(\text{C-BEL}_G(\bigwedge_{i \in G} \text{TRUST}_{\psi(z)}(i, a)) \rightarrow \text{C-BEL}_G(\psi(z)))$$

8.6.2   The Three Steps of Planning

The first phase of task division is discussion of proposals. Walton and Krabbe (1995) define this as a subtype of persuasion. This *persuasion* embedded into deliberation should result in a sequence of subgoals $\sigma = \langle \varphi_1, \ldots, \varphi_n \rangle$ and a common belief about this. During information exchange, the agents make clear their initial stand with respect to combinations of specific subgoals offered to them. Next, means-end analysis may be seen as an *inquiry* or, more commonly, another *deliberation*, matching actions α_i with subtasks φ_i. *Persuasion* and *information seeking* are then applicable during allocation of these actions, resulting in a social plan. Team members are asked, more specifically than earlier, about their abilities, opportunities and other preferences.

Even though agents are connected via the collective intention $\text{C-INT}_G(\varphi)$, being self-interested they may still have a conflict of interests during action allocation. This may call for a *negotiation* to devise a social plan reflecting the agents' individual and social interests. Finally, for the strong forms of collective commitments, announcement of *constitute*(φ, P) by a collectively trusted team member concludes action allocation, resulting in a common belief $\text{C-BEL}_G(constitute(\varphi, P))$.

Weak forms of group commitment are less demanding in terms of mental states to be achieved: the dialogues generally involve subgroups of agents and the initiator. Then it suffices that only the initiator believes the results of task division, means-end analysis and action allocation.

In any case, for all group commitments, a substantial end product of action allocation is a social plan P. Now how do agents' social commitments to their part in teamwork come into being? As a reminder, social commitments with respect to actions are defined by (see Section 4.3.2):

$$\text{COMM}(i, j, \alpha) \leftrightarrow \text{INT}(i, \alpha) \wedge \text{GOAL}(j, done(i, \alpha)) \wedge$$

$$\text{C-BEL}_{\{i,j\}}(\text{INT}(i, \alpha) \wedge \text{GOAL}(j, done(i, \alpha)))$$

For each action α from the plan, we suppose that the individual intention of one agent, say i, to execute it is in place, so $\text{INT}(i, \alpha)$. This happens usually because there is another agent, say j, who is interested for i to execute this action, so $GOAL(j, done(i, \alpha))$ is in place. Then, by communication between them, a common belief about both attitudes is created (similar to *promises* as analyzed in Searle (1969, Chapter 3)):

$$C\text{-}BEL_{\{i,j\}}(INT(i, \alpha) \land GOAL(j, done(i, \alpha)))$$

Altogether, this creates a social commitment $COMM(i, j, \alpha)$ from i to j with respect to α, as desired. In effect, all agents in the group socially commit to carry out their actions, resulting in:

$$\bigwedge_{\alpha \in P} \bigvee_{i,j \in G} COMM(i, j, \alpha).$$

Finally, a common belief:

$$C\text{-}BEL_G(\bigwedge_{\alpha \in P} \bigvee_{i,j \in G} COMM(i, j, \alpha))$$

about this is created, for example by an announcement from the initiator. This concludes the collective part of plan generation, by which the collective commitment is established.

8.6.3 *Task Division under the Magnifying Glass*

Let us turn back to a specific example: suppose a subgoal sequence $\sigma = \langle \varphi_1, \ldots \varphi_k \rangle$ needs to be found that realizes φ.

8.6.3.1 Opening

The initiator opens deliberation on task division by uttering a request to all other $i \in G$:

$$\text{request}_{a,i}(\textbf{if } division(\varphi, \sigma) \textbf{ then } \text{assert}_{i,a}(division(\varphi, \sigma))$$

$$\textbf{else } \text{assert}_{i,a}(\neg \bigvee_{\sigma \in T} division(\varphi, \sigma)))$$

Here, T is a finite set of potential subgoal sequences. Thus, if an agent knows a suitable division of φ into subgoals, it should assert this. Suppose a trusts i in everything it asserts. Then, a positive answer from i has the following consequence:

$$[\text{request}_{a,i}(\textbf{if } division(\varphi, \sigma) \textbf{ then } \text{assert}_{i,a}(division(\varphi, \sigma))$$

$$\textbf{else } \text{assert}_{i,a}(\neg \bigvee_{\sigma \in T} division(\varphi, \sigma)))]$$

$$[\text{assert}_{i,a}(division(\varphi, \sigma))]BEL(a, BEL(i, division(\varphi, \sigma)))$$

8.6.3.2   Voting

The second step consists in the initiator announcing all (or a preselected set of) answers collected from the team during the first step. Let $T_{\varphi,a} \subseteq T$ denote the finite set of candidate subgoal sequences selected by a in order to find the best sequence σ fulfilling $division(\varphi, \sigma)$. Then a divulges $T_{\varphi,a}$ to every other $i \in G$ by:

$$\text{assert}_{a,i}(\bigwedge_{\sigma \in T_{\varphi,a}} \bigvee_{i \in G} \text{BEL}(i, division(\varphi, \sigma)))$$

Subsequently, a opens the voting by requests to all $i \in G$ by announcing available options and demanding assertion of its colleagues' preferences:

$$\text{request}_{a,i}(\bigwedge_{x,y \in T_{\phi,a}} (\textbf{if}\ \ division(\varphi, x) \wedge \text{PREFER}(i, x, y)\ \textbf{then}$$

$$\text{assert}_{i,a}(\text{PREFER}(i, x, y))))$$

If i has some preferences, it responds, for example:

$$\text{assert}_{i,a}(\text{PREFER}(i, \sigma_1, \sigma_2))$$

When all voting results are in, a orders the proposed subgoal sequences σ, taking into consideration its level of trust towards particular agents in the team.

8.6.3.3   Confirming

In step three, a announces the preliminary winning subgoal sequence σ_w. Any agent having an objection should protest, as expressed by a's invitation to all other $i \in G$:

$$\text{request}_{a,i}(\textbf{if}\ \text{BEL}(i, \neg division(\varphi, \sigma_w)) \vee (\text{PREFER}(i, \sigma, \sigma_w))\ \textbf{then}$$

$$\text{assert}_{i,a}(\neg division(\varphi, \sigma_w) \vee \text{PREFER}(i, \sigma, \sigma_w)))$$

After this, a persuasion dialogue is embedded into the deliberation. If there is indeed an agent i who does not consider σ_w the best option, a may challenge it to provide a proof. Finally, after persuasion, either i admits it was wrong and the preliminary goal sequence σ_w is chosen, or i convinces the initiator that σ_w was not the best choice. In that case, a first tries to persuade the whole team of its discovery and then, if needed, a returns to the voting phase. The persuasion dialogue may have changed the team members' preferences and beliefs about the subgoal sequences and so the new voting and confirmation may have a different outcome. If still the outcome is unchanged and σ_w remains unacceptable, deliberation moves back to phase 1, as possibly agents have discovered new possibilities.

8.6.3.4   Closure

In step four, a announces the final decision, let us say for subgoal sequence σ_z by the complex speech act $\text{announce}_{a,G}(division(\varphi, \sigma_z))$, consisting of multiple repeated assertions to all other $i \in G$:

$$\text{assert}_{a,i}(division(\varphi, \sigma_z) \wedge done(\text{announce}_{a,G}(division(\varphi, \sigma_z))))$$

After these assertions, the usual consequences for other agents' belief states ensue:

$$BEL(i, division(\varphi, \sigma_z) \land done(\text{announce}_{a,G}(division(\varphi, \sigma_z))))$$

and:

$$BEL(i, BEL(a, division(\varphi, \sigma_z) \land done(\text{announce}_{a,G}(division(\varphi, \sigma_z)))))$$

Assume now, as a reasonable idealization for teamwork, that the team G commonly believes that a is trustworthy. Then we can apply the following:

$$[\text{announce}_{a,G}(division(\varphi, \sigma_z) \land done(\text{announce}_{a,G}(division(\varphi, \sigma_z))))]$$

$$(\text{C-BEL}_G(\bigwedge\nolimits_{i \in G} \text{TRUST}_{division(\varphi, \sigma_z)}(i, a)) \to \text{C-BEL}_G(division(\varphi, \sigma_z)))$$

Therefore, as desired, the deliberation results in a common belief that σ_z is an appropriate subgoal sequence realizing φ and the team can go on to start deliberation to perform means-end analysis along similar lines, resulting in a winning action sequence τ that achieves the subgoal sequence σ_z. We will not go into details here.

8.6.4 Action Allocation Under the Magnifying Glass

To conclude planning, the actions from sequence τ are assigned to team members willing and able to perform them. Action allocation is realized by information seeking, with possible phases of negotiation and persuasion. Importantly, the initiator needs to know the repertoire of actions of other agents in order to persuade them to take on certain actions when needed. First, it monitors the current situation with respect to all agents and all actions in τ. So, a asks agents $i \in G$ about their current motivational stance with respect to actions α from τ by the speech act:

$$\text{request}_{a,i}(\textbf{if } INT(i, \alpha) \textbf{ then } \text{assert}_{i,a}(INT(i, \alpha)))$$

Agents are supposed to express their intentions to perform certain actions, resulting in consequences depending on the level of trust to the initiator has in them:

$$[\text{request}_{a,i}(\textbf{if } INT(i, \alpha) \textbf{ then } \text{assert}_{i,a}(INT(i, \alpha)))]$$

$$[\text{assert}_{i,a}(INT(i, \alpha))](\text{TRUST}_\psi(a, i) \to BEL(a, INT(i, \alpha)))$$

If the initiator is also interested in i's carrying out action α, a social commitment $COMM(i, a, \alpha)$ is established. In the perfect case, all actions receive a call from a candidate i by using this method. Otherwise, for an 'orphan' action α without any bids, a may try to persuade a specific agent i to take it on, for example by convincing i that α is instrumental to achieve φ ($INSTR(i, \alpha, \varphi)$), knowing that i intends φ. Thus, the initiator utters:

$$\text{assert}_{a,i}(\bigwedge_{j \in G}(INT(j, \varphi) \to INSTR(j, \alpha, \varphi)))$$

leading to a belief change for i, who now believes that a believes its assertion:

$$[\text{assert}_{a,i}(\bigwedge_{j \in G} (\text{INT}(j, \varphi) \to \text{INSTR}(j, \alpha, \varphi)))]$$

$$\text{BEL}(i, \text{BEL}(a, \bigwedge_{j \in G}(\text{INT}(j, \varphi) \to \text{INSTR}(j, \alpha, \varphi))))$$

According to the rules, i has to either assert to have a contradictory belief or to concede. The latter option leads to a belief update for a, who now believes that i agrees with its assertion:

$$[\text{concede}_{i,a}(\bigwedge_{j \in G} (\text{INT}(j, \varphi) \to \text{INSTR}(j, \alpha, \varphi)))]$$

$$\text{BEL}(a, \text{BEL}(i, \bigwedge_{j \in G}(\text{INT}(j, \varphi) \to \text{INSTR}(j, \alpha, \varphi))))$$

Now a can combine the fact that i already intends to achieve φ with the fact that i agrees about the rule and moves to the second step, namely to persuade i that its preferred way of achieving φ is indeed by performing α. Let us adapt the definition of PREFER (see Rule (8.2)) to comparing different actions that are instrumental for the same goal:

$$\text{PREFER}(i, \alpha, \varphi) \equiv$$

$$\bigwedge_{\beta \in H} (\text{BEL}(i, \text{INSTR}(i, \beta, \varphi)) \to \text{BEL}(i, util(i, \alpha) \geq util(i, \beta)))$$

Here H is a finite pre-given set of actions. Therefore, a now utters the speech act:

$$\text{assert}_{a,i}(\bigwedge_{\beta \in H} (\text{BEL}(i, \text{INSTR}(i, \beta, \varphi)) \to \text{BEL}(i, util(i, \alpha) \geq util(i, \beta))))$$

Because during means-end analysis, i has accepted α as a part of action sequence τ realizing goal sequence σ, it is not very likely that it would question this assertion.

If it does, the challenge would have the form:

$\text{challenge}_{i,a}$

$$(\bigwedge_{\beta \in H}(\text{BEL}(i, \text{INSTR}(i, \beta, \varphi)) \to \text{BEL}(i, util(i, \alpha) \geq util(i, \beta)))) \equiv$$

$\text{assert}_{i,a}(\neg(\bigwedge_{\beta \in H}(\text{BEL}(i, \text{INSTR}(i, \beta, \varphi)) \to \text{BEL}(i, util(i, \alpha) \geq util(i, \beta)))));$

$\text{concede}_{i,a}(\text{BEL}(i, \text{INSTR}(i, \beta, \varphi)));$

$\text{request}_{i,a}(\text{assert}_{a,i}(\text{PROOF}(util(i, \alpha) \geq util(i, \beta))))$

Now a would need to provide a credible proof that indeed the utility of performing α would be better for i than that of β, which i initially prefers. If it succeeds to convince i, then a can take care of another 'orphan' action. Otherwise, it needs to move on with α to convince another agent.

For each action α that is allocated in this way, let us say to agent i, a social commitment needs to be established by a short bilateral dialogue, for example:

$$\text{assert}_{a,i}(\text{INT}(i, \alpha) \wedge \text{GOAL}(a, done(i, \alpha)))$$

immediately leading to an appropriate belief by i, who then concedes:

$$\text{concede}_{i,a}(\text{INT}(i, \alpha) \wedge \text{GOAL}(a, done(i, \alpha)))$$

Due to the fact that this short commitment protocol is commonly believed among the two agents who mutually trust each other in this matter, the concession leads to a common belief C-BEL$_{\{a,i\}}$(INT$(i, \alpha) \wedge$ GOAL$(a, done(i, \alpha))$), concluding social commitment COMM(i, a, α).

In order to complete action allocation, a collective commitment needs to be established, for example, a strong one:

$$\text{S-COMM}_{G,P}(\varphi) \leftrightarrow \text{C-INT}_G(\varphi) \wedge$$

$$constitute(\varphi, P) \wedge \text{C-BEL}_G(constitute(\varphi, P)) \wedge$$

$$\bigwedge_{\alpha \in P} \bigvee_{i,j \in G} \text{COMM}(i, j, \alpha) \wedge$$

$$\text{C-BEL}_G(\bigwedge_{\alpha \in P} \bigvee_{i,j \in G} \text{COMM}(i, j, \alpha))$$

As the collective intention is already in place, as well as a plan P and social commitments with respect to all actions from the plan, it remains for the initiator to establish common beliefs about $constitute(\varphi, P)$ and $\bigwedge_{\alpha \in P} \bigvee_{i,j \in G} \text{COMM}(i, j, \alpha)$. It can simply announce this. Assuming commonly believed trust among the team with respect to a's message, this leads to the required C-BEL$_G$($constitute(\varphi, P)$) and C-BEL$_G$($\bigwedge_{\alpha \in P} \bigvee_{i,j \in G} \text{COMM}(i, j, \alpha)$), and finally to a strong collective commitment.

In this exemplary action allocation we followed one possible scenario for a specific collective commitment. See Brzezinski *et al.* (2005) for a detailed description of an algorithm for creating a diversity of collective commitment types.

8.6.4.1 Mixing Dialogues During Planning

To sum up, a number of functional *embeddings* may take place during deliberation, possibly with subphases of inquiry and information seeking.

During *means-end analysis*, an inquiry may be embedded into deliberation, for example directly (as in Figure 8.1). More preferably, though, an elaborate deliberation including voting is performed, thus means-end analysis is carried out along similar lines as task division. In that case, the embedded dialogues may be persuasion, inquiry and information seeking.

Then, during *action allocation*, first an inquiry (possibly with its own subphases of information seeking and negotiation) is embedded into deliberation. Moreover, if necessary, negotiation (possibly with its own subphases of inquiry and information seeking)

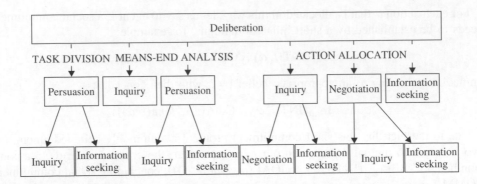

Figure 8.1 Possible dialogue embeddings during planning.

is embedded. Finally, in order to establish the pairwise social commitments, information seeking may be embedded into deliberation.

If, instead of planning from first principles, the team uses a *plan library* of more or less instantiated partial plans, input and output of planning are obviously the same as before. However, the dialogues involved are less complex here, depending on the degree of plan instantiation.

8.7 Dialogues During Team Action

Team action amounts to the execution of individual actions, usually calls for some *reconfiguration* (see Chapter 5). This complex and rather refined process demands a variety of dialogues to support cooperation and coordination.

Team action is successfully executed when all actions from the social plan P have been performed with success, leading to the achievement of goal φ. Depending on the application and the collective commitment in question, all kinds of dialogue may occur during execution of individual actions. For example, in the case of scientific collaboration (see Chapters 5 and 6 for extensive discussions), the whole team action may be viewed as *inquiry*. When team members ask the others for intermediate or final results of their individual research, *information-seeking* is embedded into inquiry. The effects of such subgroup information exchanges on the participants' individual and collective mental states are formalized similarly to information-seeking during potential recognition (see Section 8.4). *Persuasion* also occurs as a subphase of the inquiry when one agent has found a result not yet known to the others. This persuasion influences participants' mental states similarly to persuasion during team formation (see Section 8.5). Finally, by information seeking and persuasion, the agents' individual beliefs are transformed into common beliefs.

When the collective commitment is finally achieved, the initiator announces the success to the team. The strength of the resulting group belief depends on the level of trust in the initiator and can be formalized as in Section 8.5:

$$[\text{announce}_{a,G}(\text{C-COMM}_{G,P}(\varphi))]$$

$$(\bigwedge_{i \in G} \text{TRUST}(i,a) \rightarrow \bigwedge_{i \in G} \text{BEL}(i, \text{C-COMM}_{G,P}(\varphi)))$$

$$[\text{announce}_{a,G}(\text{C-COMM}_{G,P}(\varphi))]$$

$$(\text{C-BEL}_G(\bigwedge_{i \in G} \text{TRUST}(i,a)) \rightarrow \text{C-BEL}_G(\text{C-COMM}_{G,P}(\varphi)))$$

8.7.1 Communication Supports Reconfiguration

When the original collective commitment is not executed successfully, the *reconfiguration procedure* is followed. The resulting evolution includes a *revision* of motivational and informational attitudes. Pragmatically, the efficiency of revision is ensured, because agents are obliged to communicate about changes and because social commitments are present *solely* between interested partners.

Following the reconfiguration algorithm of Chapter 5, if an obstacle appears, the problems are solved by moving up in teamwork stages to the nearest point up where a different choice is possible, by depth-first search. If such a point does not exist anymore, the reconfiguration algorithm fails. This means that dialogues corresponding to the revisited stages of teamwork are relevant.

Let us trace the dialogues used during reconfiguration, focusing on the failure points of teamwork stages. Then, relevant mental attitudes are revised and information exchange stages between interested agents is performed.

The failure of potential recognition, meaning the initiator does not see any potential for cooperation towards the common goal, leads to the total failure of the system. The initiator announces this, creating an appropriate level of awareness among the team.

The failure of team formation, meaning that the collective intention cannot be established, requires a return to potential recognition to construct a new collection of potential teams. Mostly *information seeking* is involved here.

The failure of task division requires a return to team formation to establish a collective intention in the chosen new team. In this case *persuasion* is the main dialogue involved.

The failure of means-end analysis requires a return to task division in order to create a new sequence of tasks. Here, *deliberation* with possibly other embedded dialogues takes place.

The failure of action allocation requires return to means-end analysis to create a new sequence of actions corresponding to the subgoals. *Deliberation*, with possibly other embedded dialogues, mainly governs the team's reasoning about actions for goals.

Finally, during team action, after the failure of some actions from the social plan, the reasons of failure have to be recognized. Depending on the outcome, either the system fails or returns to an appropriate subphase of planning. Then again, different dialogues are applicable, as treated above.

8.8   Discussion

Related work can be found in Parsons *et al.* (1998), who also present a formal model for agent communication. They note that their own 'negotiation' in fact covers a number of Walton and Krabbe's dialogue types. Parsons *et al.* (1998) use multi-context logic in contrast to many current approaches that stick to (multi-)modal logic.

The categorization of dialogues given for human communication by Walton and Krabbe (1995) has started to inspire researchers in multi-agent systems in the 1990s, notably Chis Reed as one of the first. Reed (1998) gives a formal account of dialogue frames and functional embeddings, together with some simple examples of dialogues formulated in his formal language.

In recent years, the applications of dialogue theory in multi-agent systems have really taken off. McBurney *et al.* (2002) and Parsons *et al.* (2003) formulated important desiderata for agent argumentation protocols, including essential aspects to take into account when adapting the human-centred dialogue theory of Walton and Krabbe (1995) to multi-agent environments. Parsons *et al.* (1998) Cogan *et al.* (2005) further contributed to formalizing dialogue theory for multi-agent argumentation. An interesting formalization of deliberation was given in McBurney *et al.* (2007).

As to further research, a first issue is to investigate how the proofs given as defense of an assertion during rigorous persuasion are constructed.

Importantly, the correctness of the whole procedure of complex dialogues during teamwork is a key issue for future research. Applying contemporary methods based on automated model checkers and theorem provers seems a promising avenue in this direction.

The final question is one of balance. As argued before, even though dialogues are governed by strict rules, the reasoning needed to find the next move is highly complex. Therefore, although the result is much more flexible and refined than using a protocol like Contract Net, the process is also more time consuming. For practical applications one should carefully consider what is the priority: time or flexibility? Then one chooses the methods accordingly. See Chapter 9 for different approaches to lowering the complexity of reasoning about teamwork.

9

Complexity of Teamlog

Let your works remain a mystery.
Just show people the results.

Tao Te Ching (Lao-Tzu, Verse 36)

9.1 Computational Complexity

In this chapter, we investigate the complexity of two important subsystems of teamwork logics: TEAMLOG^{ind} and TEAMLOG. Nevertheless, the results and methods are generic and may be applied to the complexity analysis of many multi-modal logics combining interrelated agent attitudes. The results in this chapter were achieved mainly by Marcin Dziubiński (Dziubiński, 2007; Dziubiński *et al.*, 2007).

We assume that the reader is already familiar with the well-known complexity classes P, NP, co-NP, PSPACE and EXPTIME; see Balcázar *et al.* (1988) and Papadimitriou (1993). Let us only give a short informal reminder here.

- P (for *polynomial time*): decision problems which can be solved using a deterministic Turing machine within time bounded by a polynomial in the length of the input.
- NP (for *non-deterministic polynomial time*; alternative name: NPTIME): decision problems that can be solved using a non-deterministic Turing machine within time bounded by a polynomial in the length of the input. One can intuitively think of these problems in terms of 'first guessing a polynomially short witness for a positive answer; then verifying correctness for this witness in polynomial time'. A problem is NP-*complete* if it is in NP and is moreover NP-*hard*, meaning that every other problem in NP can be reduced to it in polynomial time. (Hardness and completeness are defined similarly for the other complexity classes below.) The question whether P = NP has remained the most famous open problem in complexity theory for many years (Fortnow, 2009).
- co-NP (for *complement*-NP): decision problems for which the 'No'-answers can be solved using a non-deterministic Turing machine within time bounded by a polynomial in the length of the input (corresponding to sets for which the complement is in NP). One can intuitively think of these problems in terms of 'first guessing a polynomially

short witness for a negative answer; then verifying in polynomial time that this witness indeed justifies a "No"-answer'.

- PSPACE (for *polynomial space*): decision problems that can be solved using a deterministic Turing machine within space bounded by a polynomial in the length of the input. It is well-known that PSPACE = co-PSPACE = NPSPACE = co-NPSPACE (Balcázar *et al.*, 1988; Papadimitriou, 1993).
- EXPTIME (for *exponential time*): decision problems which can be solved using a deterministic Turing machine within time exponential in the length of the input.

Roughly speaking, complexity theorists apply the term *tractable* to problems solvable with the use of reasonable resources. It is a matter of opinion whether the term 'tractable' applies to problems in P only. In recent years there has been some debate about using this term also for problems in the possibly wider class of *parameterized polynomial time*. This complexity class, called PPT, allows time polynomial in most problem parameters but exponential in a few that are known to remain small in the particular application at hand (Downey and Fellows, 1995; Hallett and Wareham, 1994; van Rooij, 2008).

Why is it important in general to investigate the exact computational complexity of a problem? Does its intractability really matter? Let us play advocate of the devil. Many real-life problems, for example in synthetic biology, are adequately solved by scientists who are blissfully oblivious of complexity theory, even though these problems are officially known to be NP-complete (Lathrop, 1994). The answer would point out that usually in these cases, the practical problems solved by synthetic biologists may be seen as specific, simpler, subcases of a known NP-complete problem. After all, demarcating a problem's complexity class as being NP-complete provides a lot of information. A solution to the P versus NP problem would have many practical consequences, favorable and unfavorable in real-life (Fortnow, 2009; van Rooij, 2008).

More specifically, in this chapter we investigate how complex it is to check *satisfiability* and *validity* of TEAMLOG formulas. Generally, why is it important for TEAMLOG, but also for other logical theories used for specifying multi-agent systems, to investigate their computational complexity? Certainly, it is quite informative to know that the satisfiability problem for TEAMLOG^{ind} is PSPACE-complete. This means that, under the supposition that P $\neq$PSPACE, it is not tractable to decide whether a certain formula follows from another. Neither is it tractable to decide whether a certain set of formulas is consistent with TEAMLOG^{ind}. This answer is significant in investigations whether a certain state of affairs can ever be reached by a given multi-agent system.

Let us turn to a short reminder about the decidability and complexity of the important questions of satisfiability, validity and model checking about logical theories in general and logics for multi-agent systems in particular.

9.1.1 Satisfiability, Validity and Model Checking

We examine the complexity of the *satisfiability problem* for our logics: given a formula φ, how much time and space (in terms of the length of φ) are needed to compute whether φ is satisfiable, that is, whether there is a suitable Kripke model $\mathcal{M}$ (from the class of structures corresponding to the logic) and a world s in it, such that $\mathcal{M}, s \models \varphi$. From this, the complexity of the *validity problem* (truth in all worlds in all suitable Kripke models) follows immediately, because φ is valid if and only if $\neg\varphi$ is not satisfiable. Consequently, if the satisfiability problem of some logic is NP-complete, then its validity problem is co-NP-complete. *Model checking*, that is evaluating truth of a given formula in a given world and model ($\mathcal{M}, s \models \varphi$), is an important related problem and is easily seen to be, at most, as complex as both satisfiability and validity. See Chapter 6 of Blackburn *et al.* (2002) for an introduction to the complexity of these problems for several standard modal logics. We do not investigate the complexity of model checking here, but see Kacprzak *et al.* (2004) and Penczek and Lomuscio (2003) for such an analysis of some MAS logics. Indeed, as long as the considered models are not too large, various contemporary methods perform model checking in a reasonable time.[1]

Unfortunately, by Cook's theorem, even satisfiability for propositional logic (called SAT) is an NP-complete problem (see Cook (1971) or Appendix C of Blackburn *et al.* (2002)). Thus, let us suppose that the famous open P and NP problem should be answered negatively and indeed P $\neq$ NP. Then modal logics interesting for multi-agent systems, all containing propositional logic as a subsystem, do not have efficiently solvable satisfiability problems.

Even though a single efficient algorithm performing well on all inputs is not possible, it is still important to discover in which complexity class a given logical theory falls. In our work we take the perspective of the system developer who wants to reason about, specify and verify a designed multi-agent system. It turns out that for many questions appearing in these processes, satisfiability tends to be easier to compute than suggested by the worst-case labels like 'PSPACE-complete' and 'EXPTIME-complete' (Halpern and Moses, 1992). It would be helpful to equip the system developer with automated, efficient tools supporting some reasoning tasks. In fact, in the discussion (Section 9.5) we will come back to methods simplifying satisfiability problems for MAS logics in an application-dependent way.

Of many single-agent modal logics with one modality, the complexity has long been known. An overview slightly extending these results is given in Halpern and Moses (1992). For us, the following results are relevant. The satisfiability problems for the systems $S5_1$ and $KD45_1$, modeling knowledge and belief of one agent, are NP-complete. Thus, perhaps surprisingly, they are no more complex than propositional logic. The complexity is increased to PSPACE if these systems are extended to more than one agent. PSPACE is also the complexity class of satisfiability for many other modal logics, for both the single- and the multi-agent cases. The basic system K_n (that we adopted for goals in Chapter 3) and the system KD_n (that we adopted for intentions in Chapter 3) can serve as examples. As soon as a notion of seemingly infinite character such as common knowledge or common belief is modeled, the complexity of the satisfiability problem jumps to EXPTIME. Intuitively, trying to find a satisfying model for a formula containing a common belief operator by the tableau method, one needs to look exponentially deep in the tableau tree to find it, while for simpler modal logics like K_n, a depth-first search through a polynomially shallow tree suffices for all formulas.[2]

---

[1] Recently, Niewiadomski *et al.* (2009) and Penczek and Szreter (2008) have developed interesting ways to translate such model checking problems into propositional logic and then apply the very fast SAT-solvers available nowadays.
[2] To be sure, having to look exponentially deep into a tableau only suggests the *risk* of EXPTIME-completeness but doesn't show it actually. For some satisfiability problems, one can combine a tableau-like method with Savitch's reachability algorithm for directed graphs to achieve a PSPACE upper bound (Savitch, 1970). For example,

9.1.2   Combination May Lead to Explosion

General results on the transfer of the complexity from single logics to their combinations are useful in investigating the complexity of multi-modal logics. Isn't satisfiability of a combination of a few PSPACE-complete logics, with some simple interdependency axioms, automatically PSPACE-complete again? Unfortunately, it turns out that the few existing positive general results (such as those in Blackburn and Spaan (1993)) apply mainly to minimal combinations, without added interdependencies, of two NP-complete systems, each with a single modality. Even more dangerously, some very negative results have been proved about the transfer of complexity to combined systems.

Most strikingly, there are two 'very decidable' logics whose combination, even without any interrelation axioms, is undecidable.

In particular for B, consider a variant of dynamic logic with two atomic programs, both deterministic. Take ; and ∩ as only operators. Satisfiability of formulas with respect to B, like that for propositional dynamic logic itself, is in EXPTIME. For C, take the logic of the global operator A (Always), defined as follows: $M, w \models A\varphi$ iff for all $v \in W$, $M, v \models \varphi$. Satisfiability for C is in NP. Blackburn and Spaan (1993) show that not only the minimal combination of B and C is *not* in EXPTIME but that it is even undecidable in any finite time (see also Blackburn *et al.* (2002), Theorem 6.1). This goes to show that one needs to be very careful with any generalization of complexity results to combined systems.

Our logic TEAMLOG and its subsystems are squarely multi-modal, not only in the sense of modeling a multi-agent version of one modal operator but also because different operators are combined and may interfere. One might expect that such a combination is much more complex than the basic multi-agent logic with one operator. In fact we show that this is not the case: TEAMLOG^{ind}, the 'individual part' of TEAMLOG, is PSPACE-complete. In order to prove this, the semantic properties relating to the interdependency axioms must be carefully translated to conditions on the multi-modal tableau with which satisfiability is tested. Of course, the challenge appears when informational and motivational group notions are added to this individual part. We show that also for this expressive system, modeling a subtle interplay between individual and group attitudes, satisfiability is EXPTIME-complete, thus of the same complexity as the system only modeling common belief. As a bonus, it turns out that even adding dynamic logic, which is relevant for our study of the attitudes' evolution in Chapter 6, does not increase complexity beyond EXPTIME.

Finally, inspired by Halpern (1995), we explore some possibilities of lowering the complexity of the satisfiability problem by restricting the modal depth of the formulas concerned or by limiting the number of propositional atoms used in the language. It turns out that bounding the depth gives a nice reduction in the individual case but is less successful where group attitudes are concerned. Combining modal depth reduction with bounding the number of propositional atoms allows for checking the satisfiability in linear time, but the constant is exponentially dependent on the number of propositions and the modal depth. Two new restrictions on modal context introduced by Dziubiński (2007)

TEAMLOG restricted to formulas of a certain type of modal context requires exponentially deep tableaus while satisfiability is still only PSPACE-complete (see Section 9.4.3).

provide a middle way: reducing complexity to PSPACE and NP, while maintaining a modicum of reasonable constants.

The rest of the chapter is structured as follows. Section 9.2 shortly reviews the language, semantics and axiom systems for the individual and group parts of our teamwork logics. In Section 9.3, the complexity of the satisfiability problem for TEAMLOG^{ind} is investigated. This is done both for the system as a whole and for some restrictions of it where formulas have bounded modal depth or the number of propositional atoms is bounded.

Section 9.4 extends the investigation to the theory TEAMLOG covering common belief and collective intention. We study both the theory as a whole and its restriction to formulas of bounded modal depth or of bounded number of propositional atoms, or allowing only restricted modal contexts. Finally, in Section 9.5 we discuss the results and present some avenues for possible extensions.

9.2   Logical Background

In most chapters of this book, we use multi-modal logics to formalize agents' informational and motivational attitudes as well as actions they perform. In this chapter, dealing with the static aspects of the agents' mental states, we only use axioms with respect to *propositions*, not actions.

Table 9.1 gathers together the formulas appearing in this chapter, together with a reminder of their intended meanings (see Chapters 2 and 3 for extensive discussions).

9.2.1   The Language

The language of TEAMLOG has been defined in Definition 3.1. For this chapter, we need to distinguish the parts of the language somewhat differently in order to separate the individual parts, as follows.

Definition 9.1 (Language)   *The languages are based on the following two sets*:

- *a countable set* $\mathcal{P}$ *of propositional symbols*;
- *a finite set* $\mathcal{A}$ *of agents, denoted by numerals* $1, 2, \ldots, n$.

Definition 9.2 (Formulas)   We inductively define the set $\mathcal{L}^{\text{ind}}$ of formulas for TEAMLOG^{ind} as follows.

Table 9.1   Formulas and their intended meanings.

BEL(i, φ)	Agent i believes that φ
E-BEL$_G(\varphi)$	Every agent in group G believes that φ
C-BEL$_G(\varphi)$	Group G has the common belief that φ
GOAL(i, φ)	Agent i has the goal to achieve φ
INT(i, φ)	Agent i has the intention to achieve φ
E-INT$_G(\varphi)$	Every agent in group G has the individual intention to achieve φ
M-INT$_G(\varphi)$	Group G has the mutual intention to achieve φ
C-INT$_G(\varphi)$	Group G has the collective intention to achieve φ

F1 each atomic proposition $p \in \mathcal{P}$ is a formula;

F2 if φ and ψ are formulas, then so are $\neg\varphi$ and $\varphi \wedge \psi$;

F3 if φ is a formula and $i \in \mathcal{A}$, then the following are formulas:

$$\text{BEL}(i, \varphi), \quad \text{GOAL}(i, \varphi), \quad \text{INT}(i, \varphi);$$

For the language $\mathcal{L}$ of TEAMLOG, an additional clause is added to the definition of formula:

F4 if φ is a formula and $G \subseteq \mathcal{A}$, then the following are formulas: E-BEL$_G(\varphi)$, C-BEL$_G(\varphi)$; E-INT$_G(\varphi)$, M-INT$_G(\varphi)$, C-INT$_G(\varphi)$.

The standard propositional constants and connectives $\top, \bot, \vee, \rightarrow$ and $\leftrightarrow$ are defined in the usual way (see Chapter 2).

9.2.2 Semantics Based on Kripke Models

Let us gather together the semantical definitions of Chapters 2 and 3 for the reader's convenience. Each Kripke model for the language $\mathcal{L}$ consists of a set of worlds, a set of accessibility relations between worlds and a valuation of the propositional atoms, as follows.

Definition 9.3 (Kripke model) *A Kripke model is a tuple* $\mathcal{M} = (W, \{B_i : i \in \mathcal{A}\}, \{G_i : i \in \mathcal{A}\}, \{I_i : i \in \mathcal{A}\}, Val)$, *such that*:

1. *W is a set of possible worlds, or states.*
2. *For all $i \in \mathcal{A}$, it holds that $B_i, G_i, I_i \subseteq W \times W$. They stand for the accessibility relations for each agent with respect to beliefs, goals and intentions, respectively.*
3. *Val : $\mathcal{P} \times W \rightarrow \{0, 1\}$ is a valuation function that assigns the truth values to atomic propositions in states.*

A Kripke frame $\mathcal{F}$ is defined as a Kripke model but without the valuation function. At this stage, it is possible to define the truth conditions pertaining to the language $\mathcal{L}$. The expression $\mathcal{M}, s \models \varphi$ is read as 'formula φ is satisfied by world s in structure $\mathcal{M}$'.

Define world t to be G_B-*reachable* (respectively G_I-*reachable*) from world s iff $(s, t) \in (\bigcup_{i \in G} B_i)^+$ (respectively $(s, t) \in (\bigcup_{i \in G} I_i)^+$). Formulated more informally, this means that there is a path of length ≥ 1 in the Kripke model from s to t along accessibility arrows B_i (respectively I_i) that are associated with members i of G.

Definition 9.4 (Truth definition) *Truth of formulas is inductively defined from Val as follows*:

- $\mathcal{M}, s \models p$ *iff* $Val(p, s) = 1$;
- $\mathcal{M}, s \models \neg\varphi$ *iff* $\mathcal{M}, s \not\models \varphi$;
- $\mathcal{M}, s \models \varphi \wedge \psi$ *iff* $\mathcal{M}, s \models \varphi$ *and* $\mathcal{M}, s \models \psi$;
- $\mathcal{M}, s \models \text{BEL}(i, \varphi)$ *iff* $\mathcal{M}, t \models \varphi$ *for all t such that sB_it*;
- $\mathcal{M}, s \models \text{GOAL}(i, \varphi)$ *iff* $\mathcal{M}, t \models \varphi$ *for all t such that sG_it*;
- $\mathcal{M}, s \models \text{INT}(i, \varphi)$ *iff* $\mathcal{M}, t \models \varphi$ *for all t such that sI_it*;

- $\mathcal{M}, s \models$ E-BEL$_G(\varphi)$ *iff for all* $i \in G$, $\mathcal{M}, s \models$ BEL(i, φ);
- $\mathcal{M}, s \models$ C-BEL$_G(\varphi)$ *iff* $\mathcal{M}, t \models \varphi$ *for all* t *that are* G_B- *reachable from* s;
- $\mathcal{M}, s \models$ E-INT$_G(\varphi)$ *iff for all* $i \in G$, $\mathcal{M}, s \models$ INT(i, φ);
- $\mathcal{M}, s \models$ M-INT$_G(\varphi)$ *iff* $\mathcal{M}, t \models \varphi$ *for all* t *that are* G_I- *reachable from* s.

In particular, this implies that for all models $\mathcal{M}$ and states s, $\mathcal{M}, s \models \top$ and $\mathcal{M}, s \not\models \bot$.

9.2.3 Axiom Systems for Individual and Collective Attitudes

For the convenience of the reader, let us give a reminder of TEAMLOG^{ind} for individual attitudes and their interdependencies, followed by our additional axioms and rules for group attitudes. These axioms and rules, together forming TEAMLOG, are fully explained in Chapters 2 and 3.

9.2.3.1 General axiom and rule

P1   All instances of propositional tautologies
PR1   From φ and $\varphi \rightarrow \psi$, derive ψ (Modus Ponens)

9.2.3.2 Axioms and rules for individual belief

For each $i \in \mathcal{A}$:

A2   BEL$(i, \varphi) \wedge$ BEL$(i, \varphi \rightarrow \psi) \rightarrow$ BEL(i, ψ) (Belief Distribution)
A4   BEL$(i, \varphi) \rightarrow$ BEL$(i, BEL(i, \varphi))$ (Positive Introspection)
A5 $\neg$BEL$(i, \varphi) \rightarrow$ BEL$(i, \neg$BEL$(i, \varphi))$ (Negative Introspection)
A6 $\neg$BEL$(i, \bot)$ (Consistency)
R2   From φ infer BEL(i, φ) (Belief Generalization)

9.2.3.3 Axioms for individual motivational operators

For each $i \in \mathcal{A}$:

A2$_D$ GOAL$(i, \varphi) \wedge$ GOAL$(i, \varphi \rightarrow \psi) \rightarrow$ GOAL(i, ψ) (Goal Distribution)
A2$_I$ INT$(i, \varphi) \wedge$ INT$(i, \varphi \rightarrow \psi) \rightarrow$ INT(i, ψ) (Intention Distribution)
R2$_D$ From φ infer GOAL(i, φ) (Goal Generalization)
R2$_I$ From φ infer INT(i, φ) (Intention Generalization)
A6$_I$ $\neg$INT$(i, \bot)$ for $i = 1, \dots, n$ (Intention Consistency)

9.2.3.4 Interdependencies between intentions and other attitudes

For each $i \in \mathcal{A}$:

A7$_{GB}$ GOAL$(i, \varphi) \rightarrow$ BEL$(i, $GOAL$(i, \varphi))$ (Positive Introspection for Goals)
A7$_{IB}$ INT$(i, \varphi) \rightarrow$ BEL$(i, INT(i, \varphi))$ (Positive Introspection for Intentions)
A8$_{GB}$ $\neg$GOAL$(i, \varphi) \rightarrow$ BEL$(i, \neg$GOAL$(i, \varphi))$ (Negative Introspection for Goals)

A8$_{IB}$ $\neg\text{INT}(i, \varphi) \rightarrow \text{BEL}(i, \neg\text{INT}(i, \varphi))$ (Negative Introspection for Intentions)
A9$_{ID}$ $\text{INT}(i, \varphi) \rightarrow \text{GOAL}(i, \varphi)$ (Intention implies Goal)

By TEAMLOG^{ind} we denote the axiom system consisting of all the above axioms and rules for individual beliefs, goals and intentions as well as their interdependencies.

9.2.3.5   Axioms and rule for general ('everyone') and common belief

C1 $\text{E-BEL}_G(\varphi) \leftrightarrow \bigwedge_{i \in G} \text{BEL}(i, \varphi)$ (General Belief)

C2 $\text{C-BEL}_G(\varphi) \leftrightarrow \text{E-BEL}_G(\varphi \wedge \text{C-BEL}_G(\varphi))$ (Common Belief)

RC1   From $\varphi \rightarrow \text{E-BEL}_G(\psi \wedge \varphi)$ infer $\varphi \rightarrow \text{C-BEL}_G(\psi)$ (Induction Rule)

9.2.3.6   Axioms and rule for general, mutual and collective intentions

M1 $\text{E-INT}_G(\varphi) \leftrightarrow \bigwedge_{i \in G} \text{INT}(i, \varphi)$ (General Intention)

M2 $\text{M-INT}_G(\varphi) \leftrightarrow \text{E-INT}_G(\varphi \wedge \text{M-INT}_G(\varphi))$ (Mutual Intention)

M3 $\text{C-INT}_G(\varphi) \leftrightarrow \text{M-INT}_G(\varphi) \wedge \text{C-BEL}_G(\text{M-INT}_G(\varphi))$ (Collective Intention)

RM1   From $\varphi \rightarrow \text{E-INT}_G(\psi \wedge \varphi)$ infer $\varphi \rightarrow \text{M-INT}_G(\psi)$ (Induction Rule)

By TEAMLOG we denote the union of TEAMLOG^{ind} with the above axioms and rules for general and common beliefs and for general, mutual and collective intentions.

Most of the axioms above, as far as they do not hold on all frames as **A2** does, correspond to well-known structural properties on Kripke frames (see Chapters 2 and Chapter 3). As a reminder, the Induction Rules **RC1** and **RM1** are sound due to the definitions of G_B-reachability and G_I-reachability in terms of the transitive closure of the union of individual relations for group G, respectively (see Definition 9.4).

9.3   Complexity of TEAMLOG^{ind}

We will show that the satisfiability problem for TEAMLOG^{ind} is PSPACE-complete. First we present an algorithm for deciding satisfiability of a TEAMLOG^{ind} formula φ working in polynomial space, thus showing that the satisfiability problem is in PSPACE. The construction of the algorithm and related results are based on the method presented in Halpern and Moses (1992). The method is centred around the well-known notions of a *propositional tableau*, a *fully expanded propositional tableau* (a set that along with any formula ψ contained in it, contains also all its subformulas, each of them either in positive or negated form), and a *tableau* designed for a particular system of multi-modal logic. Let us give adaptations of the most important definitions from Halpern and Moses (1992) as a reminder.

Definition 9.5 (Propositional tableau)   A *propositional tableau* is a set T of propositional or modal formulas such that:

- if $\neg\neg\psi \in T$ then $\psi \in T$;
- if $\varphi \wedge \psi \in T$ then both $\varphi, \psi \in T$;
- if $\neg(\varphi \wedge \psi) \in T$ then either $\neg\varphi \in T$ or $\neg\psi \in T$;
- there is no formula ψ such that ψ and $\neg\psi$ are in T.

A set of formulas T is *blatantly inconsistent* if for some formula ψ, both ψ and $\neg\psi$ are in T.

In a tableau for a modal logic, for a given formula φ, $\mathrm{Sub}(\varphi)$ denotes the set of all subformulas of φ and $\neg\mathrm{Sub}(\varphi) = \mathrm{Sub}(\varphi) \cup \{\neg\psi : \psi \in \mathrm{Sub}(\varphi)\}$.

Definition 9.6 (TEAMLOG^{ind} tableau) A TEAMLOG^{ind} tableau $\mathcal{T}$ is a tuple

$$\mathcal{T} = (W, \{B_i : i \in \mathcal{A}\}, \{G_i : i \in \mathcal{A}\}, \{I_i : i \in \mathcal{A}\}, L),$$

where W is a set of states, B_i, G_i, I_i are binary relations on W and L is a *labeling function* associating with each state $w \in W$ a set $L(w)$ of formulas, such that $L(w)$ is a propositional tableau. Here follow the two conditions that every modal tableau for our language must satisfy (Halpern and Moses, 1992):

1. If $\mathrm{BEL}(i, \varphi) \in L(w)$ and $(w, v) \in B_i$, then $\varphi \in L(v)$.
2. If $\mathrm{GOAL}(i, \varphi) \in L(w)$ and $(w, v) \in G_i$, then $\varphi \in L(v)$.
3. If $\mathrm{INT}(i, \varphi) \in L(w)$ and $(w, v) \in I_i$, then $\varphi \in L(v)$.
4. If $\neg\mathrm{BEL}(i, \varphi) \in L(w)$, then there exists a v with $(w, v) \in B_i$ and $\neg\varphi \in L(v)$.
5. If $\neg\mathrm{GOAL}(i, \varphi) \in L(w)$, then there exists a v with $(w, v) \in G_i$ and $\neg\varphi \in L(v)$.
6. If $\neg\mathrm{INT}(i, \varphi) \in L(w)$, then there exists a v with $(w, v) \in I_i$ and $\neg\varphi \in L(v)$.

Furthermore, a TEAMLOG^{ind} tableau must satisfy the following additional conditions related to axioms of TEAMLOG^{ind}:

TA6	if $\mathrm{BEL}(i, \varphi) \in L(w)$, then either $\varphi \in L(w)$ or there exists $v \in W$ such that $(w, v) \in B_i$;
TA45	if $(w, v) \in B_i$ then $\mathrm{BEL}(i, \varphi) \in L(w)$ iff $\mathrm{BEL}(i, \varphi) \in L(v)$;
TA78$_{GB}$	if $(w, v) \in B_i$ then $\mathrm{GOAL}(i, \varphi) \in L(w)$ iff $\mathrm{GOAL}(i, \varphi) \in L(v)$;
TA6$_I$	if $\mathrm{INT}(i, \varphi) \in L(w)$, then either $\varphi \in L(w)$ or there exists $v \in W$ such that $(w, v) \in I_i$;
TA78$_{IB}$	if $(w, v) \in B_i$ then $\mathrm{INT}(i, \varphi) \in L(w)$ iff $\mathrm{INT}(i, \varphi) \in L(v)$;
TA9$_{IG}$	if $(w, v) \in G_i$ and $\mathrm{INT}(i, \varphi) \in L(w)$ then $\varphi \in L(v)$.

We have the following relations between these conditions and the axioms:

- Condition **TA6** corresponds to belief consistency, axiom **A6**.[3]
- Condition **TA45** corresponds to positive and negative introspection of beliefs, axioms **A4** and **A5**.[4]

---

[3] This is a condition that occurs in Halpern and Moses (1992).

[4] We give this condition instead of two other conditions given in Halpern and Moses (1992) as correspondents to positive and negative introspection axioms in a KD45$_n$ tableau. The given condition is exactly the condition Halpern and Moses (1992) give, together with a condition corresponding to the truth axiom for S5$_n$.

- Condition **TA78$_{GB}$** corresponds to positive and negative introspection of goals, axioms **A7$_{GB}$** and **A8$_{GB}$**.
- Condition **TA78$_{IB}$** corresponds to positive and negative introspection of intentions, axioms **A7$_{IB}$** and **A8$_{IB}$**.
- Condition **TA9$_{IG}$** corresponds to the fact that intention implies goal, axiom **A9$_{IG}$**.

Given a formula φ we say that $\mathcal{T} = (W, \{B_i : i \in \mathcal{A}\}, \{G_i : i \in \mathcal{A}\}, \{I_i : i \in \mathcal{A}\}, L)$ is a TeamLog^{ind} tableau for φ if $\mathcal{T}$ is a TeamLog^{ind} tableau and there is a state $w \in W$ such that $\varphi \in L(w)$.

Throughout further discussion we will use the notion of modal depth, which we define below (the definition is for the broader language of TeamLog). Note that E-BEL and E-INT are not included in the definition, because they are viewed here as abbreviations of conjunctions of individual operators.

Definition 9.7 (Modal depth) *Let φ be a* TeamLog *formula, then* modal depth *of φ, denoted by* $\text{dep}(\varphi)$, *is defined inductively as follows*:

- $\text{dep}(p) = 0$, *where* $p \in \mathcal{P}$;
- $\text{dep}(\neg\psi) = \text{dep}(\psi)$;
- $\text{dep}(\psi_1 \text{ op } \psi_2) = \max\{\text{dep}(\psi_1), \text{dep}(\psi_2)\}$, *where* $\text{op} \in \{\wedge, \vee, \rightarrow, \leftrightarrow\}$;
- $\text{dep}(\text{OP}(i, \psi)) = \text{dep}(\psi) + 1$, *where* $\text{OP} \in \{\text{BEL, GOAL, INT}\}$;
- $\text{dep}(\text{OP}_G(\psi)) = \text{dep}(\psi) + 1$, *where* $\text{OP} \in \{\text{C-BEL, M-INT}\}$.

Let F be a set of TeamLog *formulas. Then*:

- $\text{dep}(F) = \max\{\text{dep}(\psi) : \psi \in F\}$, *if* $F \neq \varnothing$;
- $\text{dep}(\varnothing) = 0$.

The following lemma provides the equivalence between existence of a *tableau* and satisfiability for TeamLog^{ind}. It is analogous to propositions shown in Halpern and Moses (1992) for several modal logics considered there, and gives the basis for our algorithm checking TeamLog^{ind} satisfiability in Section 9.3.1.

Lemma 9.1 *A formula φ is* TeamLog^{ind} *satisfiable iff there is a* TeamLog^{ind} *tableau for φ.*

Proof The proof is very similar to the proof given in Halpern and Moses (1992) for S5$_n$ tableaus. Although we have to deal with new conditions here, this is not a problem due to the similarity of conditions **TA78$_{GB}$** and **TA78$_{IB}$** to condition **TA45**. The direction from left to right is a straightforward adaptation of the proof in Halpern and Moses (1992), and we leave it to the reader.

When constructing a model for φ out of a tableau for φ in the right to left part, we have to construct a 'serial closure' of some relations. This is done by making isolated states accessible from themselves. For example, accessibility relations I_i' for intentions would be defined on the basis of relations I_i in a tableau as follows: $I_i' = I_i'' \cup \{(w, w) : \forall v \in W(w, v) \notin I_i''\}$, where I_i'' is the smallest set containing I_i and satisfying properties corresponding to axioms **A7$_{IB}$** and **A8$_{IB}$**.

9.3.1 The Algorithm for Satisfiability of TEAMLOG^{ind}

The algorithm presented below tries to construct, for a given formula φ, a *pre-tableau*, namely, a tree-like structure that forms the basis for a TEAMLOG^{ind} tableau for φ. Nodes of this pre-tableau are labeled with subsets of $\neg\text{Sub}(\varphi)$. Nodes that are fully expanded propositional tableaus are called *states* and all other nodes are called *internal nodes*.

The algorithm is an adaptation of the algorithm presented in Halpern and Moses (1992). Modifications deal with new axioms of TEAMLOG^{ind} and corresponding properties of accessibility relations.

Input: A formula φ.

Step 1 Construct a tree consisting of single node w, with $L(w) = \{\varphi\}$.

Step 2 Repeat until none of the steps **2.1–2.3** applies:

Step 2.1 Select a leaf s of the tree such that $L(s)$ is not blatantly inconsistent and is not a propositional tableau, and select a formula ψ that violates the conditions of propositional tableau.

 Step 2.1.1 If ψ is of the form $\neg\neg\xi$ then create a successor t of s and set $L(t) = L(s) \cup \{\xi\}$.

 Step 2.1.2 If ψ is of the form $\xi_1 \wedge \xi_2$ then create a successor t of s and set $L(t) = L(s) \cup \{\xi_1, \xi_2\}$.

 Step 2.1.3 If ψ is of the form $\neg(\xi_1 \wedge \xi_2)$ then create two successors t_1 and t_2 of s and set $L(t_1) = L(s) \cup \{\neg\xi_1\}$ and $L(t_2) = L(s) \cup \{\neg\xi_2\}$.

Step 2.2 Select a leaf s of the tree such that $L(s)$ is not blatantly inconsistent and is not a fully expanded propositional tableau and select $\psi \in L(s)$ with $\xi \in \text{Sub}(\psi)$ such that $\{\xi, \neg\xi\} \cap L(s) = \varnothing$. Create two successors t_1 and t_2 of s and set $L(t_1) = L(s) \cup \{\xi\}$ and $L(t_2) = L(s) \cup \{\neg\xi\}$.

Step 2.3 Create successors of all states that are not blatantly inconsistent according to the following rules. Here, s denotes the considered state and the created successors will be called b_i-, g_i- and i_i-successors.

bel1 If $\text{BEL}(i, \psi) \in L(s)$ and there are no formulas of the form $\neg\text{BEL}(i, \chi) \in L(s)$, then let $L^{\text{BEL}_i}(s) = \{\chi : \text{BEL}(i, \chi) \in L(s)\} \cup \{\text{OP}(i, \chi) : \text{OP}(i, \chi) \in L(s)\}$, where $\text{OP} \in \{\text{BEL}, \neg\text{BEL}, \text{GOAL}, \neg\text{GOAL}, \text{INT}, \neg\text{INT}\}$. If there is no b_i-ancestor t of s, such that $L^{\text{BEL}_i}(t) = L^{\text{BEL}_i}(s)$, then create a successor u of s (called b_i-successor) with $L(u) = L^{\text{BEL}_i}(s)$.

bel2 If $\neg\text{BEL}(i, \psi) \in L(s)$, then let $L^{\neg\text{BEL}_i}(s, \psi) = \{\neg\psi\} \cup L^{\text{BEL}_i}(s)$. If there is no b_i-ancestor t of s, such that $L^{\neg\text{BEL}_i}(t, \psi) = L^{\neg\text{BEL}_i}(s, \psi)$, then create a successor u of s (called b_i-successor) with $L(u) = L^{\neg\text{BEL}_i}(s, \psi)$.

int1 If $\text{INT}(i, \psi) \in L(s)$ and there are no formulas of the form $\neg\text{INT}(i, \chi) \in L(s)$, then let $L^{\text{INT}_i}(s) = \{\chi : \text{INT}(i, \chi) \in L(s)\}$. If there is no i_i-ancestor t of s, such that $L^{\text{INT}_i}(t) = L^{\text{INT}_i}(s)$, then create a successor u of s (called i_i-successor) with $L(u) = L^{\text{BEL}_i}(s)$.

int2 If $\neg\text{INT}(i, \psi) \in L(s)$, then let $L^{\neg\text{INT}_i}(s, \psi) = \{\neg\psi\} \cup L^{\text{INT}_i}(s)$. If there is no i_i-ancestor t of s, such that $L^{\neg\text{INT}_i}(t, \psi) = L^{\neg\text{INT}_i}(s, \psi)$, then create a successor u of s (called i_i-successor) with $L(u) = L^{\neg\text{INT}_i}(s, \psi)$.

goal If $\neg\text{GOAL}(i, \psi) \in L(s)$, then let $L^{\text{GOAL}_i}(s) = \{\chi : \text{GOAL}(i, \chi) \in L(s)\}$ and $L^{\neg\text{GOAL}_i}(s, \psi) = \{\neg\psi\} \cup L^{\text{GOAL}_i}(s) \cup L^{\text{INT}_i}(s)$. If there is no g_i-ancestor t of s, such that $L^{\neg \mathrm{GOAL}_i}(t, \psi) = L^{\neg \mathrm{GOAL}_i}(s, \psi)$, then create a successor u of s (called g_i-successor) with $L(u) = L^{\neg \mathrm{GOAL}_i}(s, \psi)$.

Step 2.4 Mark a hitherto unmarked node 'satisfiable' if either it is a not blatantly inconsistent state and step **2.3** cannot be applied to it and all its successors are marked 'satisfiable' or it is an internal node having at least one descendant marked 'satisfiable'.

Step 3 If the root is marked 'satisfiable' return 'satisfiable' or otherwise return 'unsatisfiable'.

Before showing validity of the above algorithm, we will prove the following lemma which will be useful in further proofs. In what follows relations of B_i-successor, G_i-successor and I_i-successor between states will be used and are defined as follows. Let s and t be subsequent states. If t is a b_i-, g_i- or i_i-successor of some node, then t is a B_i-, G_i- or I_i-successor (respectively) of s.

Lemma 9.2 *Let s and t be states of a pre-tableau constructed by the algorithm, such that t is a B_i-successor of s and t is not blatantly inconsistent. Then the following hold for* OP $\in \{$BEL, GOAL, INT$\}$:

1. $L^{\mathrm{OP}_i}(s) = L^{\mathrm{OP}_i}(t)$.
2. $\neg \mathrm{OP}(i, \xi) \in L(s)$ *and* $L^{\neg \mathrm{OP}_i}(s, \xi) = L^{\neg \mathrm{OP}_i}(t, \xi)$, *for any* $\neg \mathrm{OP}(i, \xi) \in L(t)$.

Proof Note that if s has a B_i-successor, then it is not blatantly inconsistent. For point 1, let $\psi \in L^{\mathrm{OP}_i}(s)$.
Then we have either $\mathrm{OP}(i, \psi) \in L(s)$ (and consequently $\mathrm{OP}(i, \psi) \in L(t)$) or OP $=$ BEL, ψ is of the form BEL(i, ξ) and $\psi \in L(s)$ (consequently $\psi \in L(t)$). Thus $\psi \in L^{\mathrm{OP}_i}(t)$.

On the other hand, let $\psi \in L^{\mathrm{OP}_i}(t)$. Then either $\mathrm{OP}(i, \psi) \in L(t)$ or OP $=$ BEL, ψ is of the form BEL(i, ξ), and $\psi \in L(t)$. Suppose that the first case holds. Since $L(s)$ is a fully expanded propositional tableau, either $\mathrm{OP}(i, \psi) \in L(s)$ or $\neg \mathrm{OP}(i, \psi) \in L(s)$. Because the second possibility leads to blatant inconsistency of $L(t)$ (as by the algorithm it implies that $\neg \mathrm{OP}(i, \psi) \in L(t)$), it must be that the first possibility holds and thus $\psi \in L^{\mathrm{BEL}_i}(s)$.
The second case can be shown by similar arguments, as either BEL$(i, \xi) \in L(s)$ or BEL$(i, \xi) \in L(s)$.
For point 2, let $\neg \mathrm{OP}(i, \xi) \in L(t)$. Then by the fact that $L(s)$ is a fully expanded propositional tableau we have either $\neg \mathrm{OP}(i, \xi) \in L(s)$ or $\mathrm{OP}(i, \xi) \in L(s)$. As the second case leads to blatant inconsistency of $L(t)$, it must be the first one that holds.
$L^{\neg \mathrm{BEL}_i}(s, \xi) = L^{\neg \mathrm{BEL}_i}(t, \xi)$ can be shown by similar arguments to those used to show point 1. Note that $L^{\neg \mathrm{OP}_i}(v, \xi) = \{\neg \xi\} \cup L^{\mathrm{OP}_i}(v)$ or, in case of OP $=$ GOAL, we have $L^{\neg \mathrm{OP}_i}(v, \xi) = \{\neg \xi\} \cup L^{\mathrm{OP}_i}(v) \cup L^{\mathrm{INT}_i}(v)$.
Now we are ready to prove validity of the algorithm.

Lemma 9.3 *For any formula φ the algorithm terminates.*

Proof Let $|\varphi| = m$. For any node in a pre-tableau constructed by the algorithm, we have $| L(s) | \leq 2m$ (if $L(s)$ is not blatantly inconsistent then $| L(s) | \leq m$). Any sequence of executions of steps **2.1** and **2.2** can have length $\leq m$. Thus on the path connecting any subsequent states s and t, there can be at most $m - 1$ internal nodes.

- If s and t are states such that t is a G_i-successor or I_i-successor of s, then $\mathrm{dep}(L(t)) < \mathrm{dep}(L(s))$.
- If t is a B_i-successor of s and u is a B_j-successor of t, where $i \neq j$, then $\mathrm{dep}(L(u)) < \mathrm{dep}(L(s))$.
- If t is a B_i-successor of s then, by Lemma 9.2, t cannot have any B_i-, G_i-nor I_i-successors. Thus, for any successor node u of t, $\mathrm{dep}(L(s)) < \mathrm{dep}(L(u))$.

All of the above arguments show that a pre-tableau constructed by the algorithm can have a depth at most $2 \cdot \mathrm{dep}(\varphi)m$. Since $\mathrm{dep}(\varphi) \leq m - 1$, the modal depth of a pre-tableau is bounded by $2 \cdot m(m - 1)$. This also shows that the algorithm terminates.

Lemma 9.4 *A formula φ is satisfiable iff the algorithm returns 'satisfiable' on input φ.*

Proof For the right to left direction, a tableau

$$\mathcal{T} = (W, \{B_i : i \in \mathcal{A}\}, \{G_i : i \in \mathcal{A}\}, \{I_i : i \in \mathcal{A}\}, L)$$

based on the pre-tableau is constructed by the algorithm. W is the set of states of the pre-tableau. For $\{w, v\} \subseteq W$, let $(w, v) \in B_i'$ if v is the closest descendant state of w and the first successor of w on the path between w and v is a b_i-successor of w. Then B_i is determined as the transitive Euclidean closure of the above relation B_i'.

Relations G_i and I_i are defined analogously, but without taking the transitive Euclidean closure. Labels of states in W are the same as in the pre-tableau. Checking that $\mathcal{T}$ is a $\mathrm{TEAMLOG}^{\mathrm{ind}}$ tableau is very much like in the case of $S5_n$ tableaus, with the new conditions **TA6**, **TA45**, **TA78**$_{GB}$, **TA6**$_I$ and **TA78**$_{IB}$ being the most difficult cases.

For **TA6**, note that if $v \in W$ has no successor states and $\mathrm{BEL}(i, \psi) \in L(v)$, then v cannot be a root, otherwise there is no ancestor of v such that its label is $L^{\mathrm{BEL}_i}(v)$, so step **2.3. case bel1** of the algorithm applies to v and it cannot be a leaf. Therefore, there is a $w \in W$, such that $(w, v) \in B_i$.

Since $\mathrm{BEL}(i, \psi)$ is a subformula of φ, then either $\neg\mathrm{BEL}(i, \psi) \in L(w)$ or $\mathrm{BEL}(i, \psi) \in L(w)$. Because the first possibility leads to contradiction with $\mathrm{BEL}(i, \psi) \in L(v)$, then it must be the second, and this implies $\psi \in L(v)$.

Condition **TA**$_I$ can be shown similarly.

Condition **TA45** is also based on the fact that labels of states are fully expanded propositional tableaus, and can be shown similarly to **TA6** (Halpern and Moses, 1992). Since **TA78**$_{GB}$ and **TA78**$_{IB}$ are very similar to **TA45**, a b_i-successor inherits all formulas of the form $\mathrm{GOAL}(i, \psi)$, $\neg\mathrm{GOAL}(i, \psi)$, $\mathrm{INT}(i, \psi)$ and $\neg\mathrm{INT}(i, \psi)$, thus they can be shown analogously to **TA45**. Lemma 9.1 gives the final result.

For the left to right direction we show, for any node w in the pre-tableau, the claim that if w is not marked 'satisfiable' then $L(w)$ is inconsistent. From this it follows that if the root is not marked 'satisfiable' then $\neg\varphi$ is provable and thus φ is unsatisfiable.

The claim is shown by induction on the length of the longest path from a node w to a leaf of the pre-tableau. Most cases are easy and can be shown similarly to the case of $S5_n$ presented in Halpern and Moses (1992). We show only the most difficult case connected with new axioms of $\mathrm{TEAMLOG}^{\mathrm{ind}}$, namely the one in which w is not a leaf and has a b_i-successor v generated by a formula of the form $\mathrm{BEL}(i, \psi) \in L(w)$ (other cases are either similar or easier). Since by induction hypothesis $L(v)$ is inconsistent, we can show using **A2**, **R1** and **R2** that the set

$$X = \qquad \{\mathrm{BEL}(i, \psi) : \mathrm{BEL}(i, \psi) \in L(w)\} \cup$$

$$\{\mathrm{BEL}(i, \psi) : \psi \in L(w) \text{ and is of the form } \mathrm{OP}(i, \chi)\}$$

proves $\mathrm{BEL}(i, \bot)$, so by **A6**, X is also inconsistent. Assume that $L(w)$ is consistent, then the set

$$Y = L(w) \cup \{\mathrm{BEL}(i, \psi) : \psi \in L(w) \text{ and is of the form } \mathrm{OP}(i, \chi)\} \cup \{\neg\mathrm{BEL}(i, \bot)\}$$

is also consistent (by axioms **A4-6**, **A7-8**$_{GB}$ and **A7-8**$_{IB}$). This leads to contradiction, since $X \subseteq Y$, and thus $L(w)$ must be inconsistent.

Theorem 9.1 *The satisfiability problem for* TEAMLOG^{ind} *is PSPACE-complete.*

Proof Since the depth of the pre-tableau constructed by the algorithm for a given φ is at most $2 \cdot |\varphi|(|\varphi| - 1)$ and the algorithm is deterministic, it can be run on a deterministic Tuning machine by depth-first search using polynomial space. Thus TEAMLOG^{ind} is in PSPACE. On the other hand the problem of KD$_n$ satisfiability, known to be PSPACE-hard, can be reduced to TEAMLOG^{ind} satisfiability, so TEAMLOG^{ind} is PSPACE-complete.[5]

9.3.2   *Effect of Bounding Modal Depth for* TEAMLOG^{ind}

As was shown in Halpern (1995), bounding the modal depth of formulas by a constant results in reducing the complexity of the satisfiability problem for modal logics K$_n$, KD$_n$ and KD45$_n$ to NP-complete.[6] An analogous result holds for the logic TEAMLOG^{ind}, as we shall now show.

Theorem 9.2 *For any fixed k, if the set of propositional atoms $\mathcal{P}$ is infinite and the modal depth of formulas is bounded by k, then the satisfiability problem for* TEAMLOG^{ind} *is NP-complete.*

Proof From the proof of **Lemma 9.3** we can observe that the number of states on a path from the root of a pre-tableau constructed by the algorithm to a leaf depends linearly on the modal depth of the input formula (it is $\leq 2 \cdot \mathrm{dep}(\varphi)$). Thus the size of the tableau corresponding to this pre-tableau is bounded by $O(|\varphi|^{2 \cdot \mathrm{dep}(\varphi)})$. This means that the satisfiability of the formula φ with bounded modal depth can be checked by the following non-deterministic algorithm.

Input: A formula φ.
Step 1 Guess a tableau $\mathcal{T}$ satisfying φ.
Step 2 Check that $\mathcal{T}$ is indeed a tableau for φ.

Since the tableau $\mathcal{T}$ constructed at step **1** of the algorithm is of polynomial size, step **2** can be realized in polynomial time. Thus the satisfiability problem of TEAMLOG^{ind}-formulas with modal depth bounded by a constant is in NP-time. It is also NP-complete,

---

[5] Recall that even KD (for one agent) is PSPACE-complete.
[6] Actually, in Halpern (1995) the logic T$_n$ (not KD$_n$) is considered but all proofs there that work for T$_n$ work also for KD$_n$.

as the satisfiability problem for propositional logic is NP-hard and propositional logic is included in TEAMLOG^{ind} for formulas with bounded modal depth.

9.3.3 Effect of Bounding the Number of Propositional Atoms for TEAMLOG^{ind}

Another natural constraint on the language is bounding the number of propositional atoms. As was shown in Halpern (1995), constraining the language of the logics K_n, KD_n (for $n \geq 1$) and $KD45_n$ (for $n \geq 2$) this way does not change the hardness of the satisfiability problem for them, even if $|\mathcal{P}| = 1$. This result holds also for our logic, as the formula used in the proof of that fact in Halpern (1995) could be expressed in TEAMLOG^{ind} with the use of the INT modality.

Similarly to Halpern (1995), we can show that if bounding the number of propositional atoms is combined with bounding the modal depth of formulas, the complexity is reduced to linear time.

Theorem 9.3 *For any fixed $k, l \geq 1$, if the number of propositional atoms is bounded by l and the modal depth of formulas is bounded by k, then the satisfiability problem for TEAMLOG^{ind} can be solved in linear time.*

Proof By the same argument as in Halpern (1995), if $|\mathcal{P}| \leq l$, then there is only a finite number of equivalence classes (based on logical equivalence) of formulas of modal depth bounded by k in the language of TEAMLOG^{ind}. This can be proved by induction on k (see for example Blackburn *et al.* (2002), Proposition 2.29). Thus there is a finite set $\varphi_1, \ldots, \varphi_N$ of satisfiable formulas, each witness of a particular equivalence class all of whose members are satisfiable, and a corresponding fixed finite set of models $M_1, \ldots, M_N$ satisfying these formulas.

To check the satisfiability of a formula, it is enough to check whether it is satisfied in one of these models $M_1, \ldots, M_N$, and this can be done in time linear in the length of the formula; as the set of relevant models is fixed, it only contributes to the constant factor.

9.4   Complexity of the System TEAMLOG

We will show that the satisfiability problem for the system TEAMLOG, including group notions like common belief and mutual intentions, is EXPTIME-complete. First we prove that TEAMLOG has the small model property in the sense that for each satisfiable formula φ, a satisfying model of size $\mathcal{O}(2^{|\varphi|})$ can be found. To show this, a filtration technique is used (Blackburn *et al.*, 2002). Let $G \subseteq \{1, \ldots, n\}$.

Definition 9.8 *A set of formulas Σ which is closed for subformulas is* closed *if it satisfies the following*:

Cl1   *if* C-BEL$_G(\varphi) \in \Sigma$, *then* E-BEL$_G(\varphi \wedge$ C-BEL$_G(\varphi)) \in \Sigma$;
Cl2   *if* E-BEL$_G(\varphi) \in \Sigma$, *then* $\{$BEL$(j, \varphi) : j \in G\} \subseteq \Sigma$;
Cl3   *if* M-INT$_G(\varphi) \in \Sigma$, *then* E-INT$_G(\varphi \wedge$ M-INT$_G(\varphi)) \in \Sigma$;
Cl4   *if* E-INT$_G(\varphi) \in \Sigma$, *then* $\{$INT$(j, \varphi) : j \in G\} \subseteq \Sigma$.

Let $\mathcal{M} = (W, \{B_i : i \in \mathcal{A}\}, \{G_i : i \in \mathcal{A}\}, \{I_i : i \in \mathcal{A}\}, Val)$ be a TEAMLOG^{ind} model, Σ a closed set and let $\equiv_f^\Sigma \subseteq \Sigma \times \Sigma$ be an equivalence relation such that, for $\{w, v\} \subseteq W$, $w \equiv_f^\Sigma v$ iff for any $\varphi \in \Sigma$, $\mathcal{M}, w \models \varphi \Leftrightarrow \mathcal{M}, v \models \varphi$. As usual, $[w]$ denotes the equivalence class of $[w]$. Let

$$\mathcal{M}_\Sigma^f = (W^f, \{B_i^f : i \in \mathcal{A}\}, \{G_i^f : i \in \mathcal{A}\}, \{I_i^f : i \in \mathcal{A}\}, Val^f)$$

be defined as follows:

F0 $W^f = W/_{\equiv_f^\Sigma}$, $Val^f(p, [w]) = Val(p, w)$.

F1 $B_i^f = \{([w], [v]) : \text{for any } \text{BEL}(i, \varphi) \in \Sigma, \ \mathcal{M}, w \models \text{BEL}(i, \varphi) \Rightarrow \mathcal{M}, v \models \varphi$ and for any $\text{OP}(i, \varphi) \in \Sigma, \ \mathcal{M}, w \models \text{OP}(i, \varphi) \Leftrightarrow \mathcal{M}, v \models \text{OP}(i, \varphi)\}$, where $\text{OP} \in \{\text{BEL},$ $\text{GOAL, INT}\}$.

F2 $G_i^f = \{([w], [v]) : \text{for any } \text{GOAL}(i, \varphi) \in \Sigma, \ \mathcal{M}, w \models \text{GOAL}(i, \varphi) \Rightarrow \mathcal{M}, v \models \varphi$ and for any $\text{INT}(i, \varphi) \in \Sigma, \ \mathcal{M}, w \models \text{INT}(i, \varphi) \Rightarrow \mathcal{M}, v \models \varphi\}$.

F3 $I_i^f = \{([w], [v]) : \text{for any } \text{INT}(i, \varphi) \in \Sigma, \ \mathcal{M}, w \models \text{INT}(i, \varphi) \Rightarrow \mathcal{M}, v \models \varphi\}$.

It is easy to check that if $\mathcal{M}$ is a TEAMLOG^{ind} model, then so is $\mathcal{M}_\Sigma^f$ and, moreover, that if Σ is a closed set, then $\mathcal{M}_\Sigma^f$ is a *filtration* of $\mathcal{M}$ through Σ. This leads to the following standard lemma (thus left without a proof, but see Blackburn *et al.* (2002)):

Lemma 9.5 *If $\mathcal{M}$ is a TEAMLOG^{ind} model and Σ is a closed set of formulas, then for all $\varphi \in \Sigma$ and all $w \in W$, $\mathcal{M}, w \models \varphi$ iff $\mathcal{M}_\Sigma^f, [w] \models \varphi$.*

From Lemma 9.5 it follows that TEAMLOG has the finite model property and that its satisfiability problem is decidable. Let $\text{Cl}(\varphi)$ denote the smallest closed set containing $\text{Sub}(\varphi)$, and let $\neg \text{Cl}(\varphi)$ consist of all formulas in $\text{Cl}(\varphi)$ and their negations. If a formula φ is satisfiable then it is satisfiable in a filtration through $\text{Cl}(\varphi)$, and any such filtration has at most $|P(\text{Cl}(\varphi))| = \mathcal{O}(2^{|\varphi|})$ states.

Now we present an exponential time algorithm for checking TEAMLOG satisfiability of a formula φ. The algorithm and the proof of its validity are modified versions of the algorithm for checking satisfiability for propositional dynamic logic (PDL) and its validity proof presented in Harel *et al.* (2000)[7]. The algorithm attempts to construct a model $\mathcal{M} = \mathcal{N}_{\text{Cl}(\varphi)}^f$, where $\mathcal{N}$ is a *canonical model* for TEAMLOG. This is done by constructing a sequence of models $\mathcal{M}^k$, being subsequent approximations of $\mathcal{M}$ as follows.

Input: A formula φ
Step 1 Construct a model

$$\mathcal{M}0 = (W0, \{B0_i : i \in \mathcal{A}\}, \{G0_i : i \in \mathcal{A}\}, \{I0_i : i \in \mathcal{A}\}, Val^0),$$

where $W0$ is the set of all maximal subsets of $\neg \text{Cl}(\varphi)$, that is sets that for every $\psi \in \text{Cl}(\varphi)$ contain either ψ or $\neg \psi$, $Val^0(p, w) = 1$ iff $p \in w$ and accessibility relations

---

[7] Note that one can see TEAMLOG as a modified and restricted version of PDL, where the BEL, GOAL and INT operators for each agent are seen as atomic programs satisfying some additional axioms, while group operators can be defined as complex programs using the $\cup$ and $*$ operators.

are defined analogously as in $\mathcal{M}^f_{Cl(\varphi)}$. We present the definition of $B0_i$, which makes definitions **G1**, **I1** of $G0_i$ and $I0_i$ obvious:

B1

$$BO_i = \{(w, v) : \text{for any } BEL(i, \varphi) \in Cl(\varphi), BEL(i, \varphi) \in w \Rightarrow \varphi \in v$$

$$\text{and for any } OP(i, \varphi) \in Cl(\varphi), \ OP(i, \varphi) \in w \Leftrightarrow OP(i, \varphi) \in v\}$$

where $OP \in \{BEL, GOAL, INT\}$.

Step 2 Construct a model $\mathcal{M}1$ by removing from $W0$ states that are not closed propositional tableaus.

Step 3 Repeat the following, starting with $k = 0$, until no state can be removed.

Step 3.1 Find a formula $\psi \in \neg Cl(\varphi)$ and state $w \in W^k$ such that $\psi \in w$ and one of the conditions below is not satisfied. If such a state was found, remove it from W^k to obtain W^{k+1}.

AB1 if $\psi = \neg BEL(i, \chi)$, then there exists $v \in B^k_i$ such that $\neg\chi \in v$;

AG1 if $\psi = \neg GOAL(i, \chi)$, then there exists $v \in G^k_i$ such that $\neg\chi \in v$;

AI1 if $\psi = \neg INT(i, \chi)$, then there exists $v \in I^k_i$ such that $\neg\chi \in v$;

AB2 if $\psi = BEL(i, \chi)$, then there exists $v \in B^k_i$ such that $\chi \in v$;[8]

AI2 if $\psi = INT(i, \chi)$, then there exists $v \in I^k_i$ such that $\chi \in v$;

AEB1 if $\psi = \neg E\text{-}BEL_G(i, \chi)$, then there exists $v \in B^k_G$ (where $B^k_G = \bigcup_{j \in G} B^k_j$) such that $\neg\chi \in v$;

AEI1 if $\psi = \neg E\text{-}INT_G(i, \chi)$, then there exists $v \in I^k_G$ (where $I^k_G = \bigcup_{j \in G} I^k_j$) such that $\neg\chi \in v$;

ACB1 if $\psi = \neg C\text{-}BEL_G(i, \chi)$, then there exists $v \in (B^k_G)^+$ such that $\neg\chi \in v$;

AMI1 if $\psi = \neg M\text{-}INT_G(i, \chi)$, then there exists $v \in (I^k_G)^+$ such that $\neg\chi \in v$.

Step 4 If there is a state in the model $\mathcal{M}^l$ obtained after step **3** containing φ, then return 'satisfiable', otherwise return 'unsatisfiable'.

It is obvious that the algorithm terminates. Moreover, since each step can be done in polynomial time, the algorithm terminates after $\mathcal{O}(2^{|\varphi|})$ steps. To prove the validity of the algorithm, we have to prove an analogue to a lemma in Harel *et al.* (2000). In the following lemma, $OP_G \in \{E\text{-}BEL_G, E\text{-}INT_G\}$, $OP^+_G \in \{C\text{-}BEL_G, M\text{-}INT_G\}$ and R denotes the relation corresponding to operator OP used in the particular context.

Lemma 9.6 *Let $k \geq 1$ and assume that $\mathcal{M} \subseteq \mathcal{M}^k$. Let $\chi \in Cl(\varphi)$ be such that every formula from $Cl(\chi)$ of the form $OP(i, \psi)$, $OP_G(\psi)$ or $OP^+_G(\psi)$ and $w \in W^k$ satisfies the conditions of step **3** of the algorithm. Then:*

1 *for all $\xi \in Cl(\chi)$ and $v \in W^k$, $\xi \in v$ iff $\mathcal{M}, v \models \xi$.*

2.1 *for any $OP(i, \xi) \in Cl(\chi)$ and $\{w, v\} \subseteq W^k$:*

    2.1.a *if $(w, v) \in R_i$ then $(w, v) \in R^k_i$;*

    2.1.b *if $(w, v) \in R^k_i$ and $OP(i, \xi) \in v$ then $\xi \in v$.*

---

[8] The conditions in Step 3.1 are analogous to conditions for PDL. The only differences are conditions **AB2** and **AI2** that correspond to axioms **A6** and **A6$_I$** and that ensure that all worlds have belief- and intention-successors.

2.2 *for any* $OP_G(\xi) \in Cl(\chi)$ *and* $\{w, v\} \subseteq W^k$:
    2.2.a *if* $(w, v) \in R_G$ *then* $(w, v) \in R_G^k$;
    2.2.b *if* $(w, v) \in R_G^k$ *and* $OP_G(\xi) \in v$ *then* $\xi \in v$.
2.3 *for any* $OP_G^+(\xi) \in Cl(\chi)$ *and* $\{w, v\} \subseteq W^k$:
    2.3.a *if* $(w, v) \in (R_G)^+$ *then* $(w, v) \in (R_G^k)^+$;
    2.3.b *if* $(w, v) \in (R_G^k)^+$ *and* $OP_G^+(\xi) \in v$ *then* $\xi \in v$.

Proof The proof is analogous to the one of the lemma for PDL in Harel *et al.*, (2000) and the additional properties of TEAMLOG do not affect the argumentation. The proof of points **2.1–2.3** is essentially based on the fact that $\mathcal{M}$ is a filtration and similar techniques are used here to those from the proof of the filtration lemma. The proof of point **1** is by induction on the structure of ξ, similarly to its analogue for the lemma for PDL.

Lemma 9.7 *A formula φ is satisfiable iff the algorithm returns 'satisfiable' on input φ.*

Proof Since every state $w \in W$ is a maximal subset of $\neg Cl(\varphi)$, we have $W \subseteq W0$. Moreover, since every state $w \in W$ is a propositional tableau satisfying conditions from step **2** of the algorithm, therefore $W \subseteq W1$. Conditions in step **2** also guarantee that no state $w \in W$ can be deleted in step **3**. This shows that $W \subseteq W^k$, for all W^k constructed throughout an execution of the algorithm. It follows that if φ is satisfiable, then the algorithm will return 'satisfiable'.

If model $\mathcal{M}^l$ obtained after step **3** of the algorithm is not empty, then it can be easily checked that it is a TEAMLOG^{ind} model. This is because every model $\mathcal{M}^k$ constructed throughout an execution of the algorithm preserves conditions **B1, G1, I1**. Moreover, conditions **AB2, AI2** guarantee that the relations B_i^l and I_i^l are serial. Now, if there is a $w \in W^l$ such that $\varphi \in w$, then (by **1** of Lemma 9.7) $\mathcal{M}^l, w \models \varphi$. Since $\mathcal{M}^l$ is a TEAMLOG^{ind} model, then φ is TEAMLOG satisfiable. So the algorithm is valid.

Theorem 9.4 *The satisfiability problem for TEAMLOG is EXPTIME-complete.*

Proof Immediately from Lemmas 9.5, 9.6, and 9.7, it follows that satisfiability is in EXPTIME. It is also EXPTIME-hard by the same proof as used in Theorem 9.5 in the next section.

Remark 9.1 *The algorithm above and Lemma 9.6. are kept similar to the ones presented in Harel et al. (2000), so one can combine them to obtain a deterministic exponential time algorithm for a combination of TEAMLOG and PDL.*

9.4.1 Effect of Bounding Modal Depth for TEAMLOG

The effect of bounding the modal depth of formulas on the complexity of the satisfiability problem for TEAMLOG is not as promising as in the case of TEAMLOG^{ind}. It can be shown that even if modal depth is bounded by 2, the satisfiability problem remains EXPTIME-hard. The proof we give here is inspired by the proof of EXPTIME-hardness of the satisfiability problem for PDL given in Blackburn *et al.* (2002, Chapter 6.8).

Theorem 9.5 *The satisfiability problem for deciding satisfiability of TEAMLOG formulas with modal depth bounded by 2 is EXPTIME-complete.*

Proof The fact that satisfiablity for formulas of modal depth bounded by 2 is in EXP-TIME follows immediately by Theorem 9.4, as a special case. Thus, what we need to show is EXPTIME-hardness of the problem. To do this, we will use the two-person corridor tiling game.

A tile is a 1×1 square, with fixed orientation and a color assigned to each side. There are two players taking part in the game and a referee who starts the game. The referee gives the players a finite set $\{T_1, \ldots, T_s\}$ of tile types. Players will use tiles of these types to arrange them on the grid in such a way that the colors on the common sides of adjacent tiles match. Additionally there are two special tile types T_0 and T_{s+1}. T_0 is an all sides white type, used merely to mark the boundaries of the corridor inside which the two players will place their tiles. T_{s+1} is a special winning tile that can be placed only in the first column.

At the start of the game, the referee fills in the first row (places $\{1, \ldots, m\}$) of the corridor with m initial tiles of types $\{T_1, \ldots, T_s\}$ and places two columns of T_0 type tiles in columns 0 and $m + 1$ marking the boundaries of the corridor. Now the two players A and B place their tiles in alternating moves. Player A is the one to start. The corridor is to be filled row by row from bottom to top and from left to right. Thus the place of the next tile is determined and the only choice the players make is the type of tile to place. The color of a newly placed tile must fit the colors of its adjacent tiles. We will use $C(T', T, T'')$ to denote that T can be placed to the right of T' and above tile T'', thus that $right(T') = left(T)$ and $top(T'') = bottom(T)$, where $right$, $left$, top and $bottom$ give the colors of respective sides of a tile.

If after finitely many rounds a tiling is constructed in which a tile of type T_{s+1} is placed in the first column, then player A wins. Otherwise, that is if no player can make a legal move or if the game goes on infinitely long and no tile of type T_{s+1} is placed in the first column, player B wins. The problem of deciding if for a given setting of the game there is a winning strategy for player A is an EXPTIME-hard problem (Chlebus, 1986). Following Blackburn *et al.* (2002, Chapter 6.8) we will show that this problem can be reduced to the satisfiability problem of TEAMLOG formulas of modal depth ≤ 2.

In the proof of Blackburn *et al.* (2002, Chapter 6.8) a formula is constructed for a given tiling game, such that a model of it is the game tree for given settings of the game with its root as a current state. States of the tree contain information about the actual configuration of the tiles, the player who is to move next and the position at which the next tile is to be placed. The depth of the tree is bounded by m^{s+2}. Note that after m^{s+2} rounds, repetition of rows must have occurred and if A can win a game with repetitions, A can also win a game without them, thus it is enough to consider m^{s+2} rounds only.

The formula from the proof of Blackburn *et al.* (2002, Chapter 6.8) uses two PDL modalities $[a]$ and $[a^*]$ and its depth is bounded by 2. These modalities could be replaced by INT$(1, \cdot)$ and M-INT$'_{\{1\}}$, where M-INT$'_G(\varphi)$ is a shortcut for M-INT$'_G(\varphi) \wedge \varphi$ (recall that $[a^*]$ is reflexive and M-INT is not). The proof would remain the same. Thus it can be shown that even if we consider M-INT with $n \geq 1$ and formulas with modal depth bounded by 2, the satisfiability problem remains in EXPTIME. Below we show a slightly modified version of the Blackburn *et al.* (2002, Chapter 6.8) proof, adapted for C-BEL. In this case $n \geq 2$ is required. This is not surprising, as for $n = 1$, C-BEL is equivalent to BEL because by axioms **A4** and **A5**, BEL$(1, \varphi)$ and BEL$(1, \text{BEL}(1, \varphi))$ are equivalent.

Let $\mathcal{G} = (m, \mathcal{T}, (I_1, \ldots, I_m))$, where $\mathcal{T} = \{T_0, \ldots, T_{s+1}\}$ and $I_j \in \mathcal{T}$ for $0 \le j \le m$, be a setting for a two person corridor tiling game described above. Here, $(I_1, \ldots, I_m)$ is the row of types of the initial tiling of the first row of the corridor. We construct a formula $\varphi(\mathcal{G})$ such that it is satisfiable iff player A has a winning strategy. The following propositional symbols are used to construct a formula:

- a to indicate that A has the next move; we will also use p_1 to denote a and p_2 to denote $\neg a$ in order to shorten some formulas;
- $pos_1, \ldots, pos_m$ to indicate the column in which a tile is to be placed in the current round;
- $col_i(T)$, for $0 \le i \le m + 1$ and $T \in \mathcal{T}$, to indicate that a tile previously placed in column i is of type T;[9]
- win to indicate that the current position is a winning position for A;
- $q_1, \ldots q_N$, where $N = \lceil \log_2 (m^{s+2}) \rceil$, to enumerate states. Boolean values of these variables in a given state can be treated as a representation of a binary number with q_1 being the least significant bit and q_N being the most significant one. We will give the same number to all states belonging to the same round and we will use the notation $round = k$ as a shortcut for the formula expressing that the number encoded by $q_N \ldots q_1$ is equal to k.

The formula $\varphi(\mathcal{G})$ will be composed of the following formulas describing settings of the game and giving necessary and sufficient conditions for the existence of a winning strategy for A. In what follows, $k \in \{1, 2\}$, $0 \le i \ne j \le m + 1$, $0 \le x \ne y \le s + 1$ and $\{T, T', T''\} \subseteq \mathcal{T}$ (if not stated differently). We will also use C-BEL$'_G(\varphi)$ as a shortcut for C-BEL$_G(\varphi) \wedge \varphi$. We also use the standard conventions that $\bigwedge \varnothing = \top$ and $\bigvee \varnothing = \bot$.

$$a \wedge pos_1 \wedge col_0(T_0) \wedge col_{m+1}(T_0) \wedge col_1(I_1) \wedge \ldots \wedge col_m(I_m) \tag{9.1}$$

$$\text{C-BEL}_{\{1,2\}}(pos_1 \vee \ldots \vee pos_m) \tag{9.2}$$

$$\text{C-BEL}'_{\{1,2\}}(pos_i \rightarrow \neg pos_j), \; 1 \le i \ne j \le m \tag{9.3}$$

$$\text{C-BEL}_{\{1,2\}}(col_i(T_0) \vee \ldots \vee col_i(T_{s+1})) \tag{9.4}$$

$$\text{C-BEL}'_{\{1,2\}}(col_i(T_x) \rightarrow \neg col_i(T_y)) \tag{9.5}$$

$$\text{C-BEL}_{\{1,2\}}(col_0(T_0) \wedge col_{m+1}(T_0)) \tag{9.6}$$

$$\text{C-BEL}'_{\{1,2\}}(\neg pos_i \rightarrow ((col_i(T_x) \rightarrow \text{BEL}(k, col_i(T_x))) \wedge$$
$$(\neg col_i(T_x) \rightarrow \text{BEL}(k, \neg col_i(T_x)))))) \tag{9.7}$$

$$\text{C-BEL}'_{\{1,2\}}((pos_m \wedge p_k \rightarrow \text{BEL}(k, pos_1)) \wedge$$
$$(pos_1 \wedge p_k \rightarrow \text{BEL}(k, pos_2)) \wedge \ldots \wedge (pos_{m-1} \wedge p_k \rightarrow \text{BEL}(k, pos_m))) \tag{9.8}$$

$$\text{C-BEL}'_{\{1,2\}}((a \rightarrow \text{BEL}(1, \neg a)) \wedge (\neg a \rightarrow \text{BEL}(2, a))) \tag{9.9}$$

$$\text{C-BEL}'_{\{1,2\}}\Big(pos_i \wedge col_{i-1}(T') \wedge col_i(T'') \wedge p_k \rightarrow$$

---

[9] Note that $col_i(T)$ is a parametrized name of a propositional symbol.

$$\text{BEL}\left(k, \bigvee\{col_i(T) : C(T', T, T'')\}\right)\right),\ 1 \leq i \leq m \tag{9.10}$$

$$\text{C-BEL}'_{\{1,2\}}\left(pos_n \rightarrow \text{BEL}\left(k, \bigvee\{col_n(T) : right(T) = white\}\right)\right) \tag{9.11}$$

$$\text{C-BEL}'_{\{1,2\}}\left(\neg a \wedge pos_i \wedge col_i(T'') \wedge col_{i-1}(t') \rightarrow\right.$$

$$\left.\bigwedge\{\neg \text{BEL}(k, \neg col_i(T)) : C(T', T, T'')\}\right),\ 1 \leq i \leq m \tag{9.12}$$

$$win \wedge \text{C-BEL}'_{\{1,2\}}(win \rightarrow (col_1(T_{s+1}) \vee (a \wedge \neg \text{BEL}(1, \neg win)) \vee$$

$$(\neg a \wedge \text{BEL}(2, win)))) \tag{9.13}$$

$$\text{C-BEL}'_{\{1,2\}}((round = N) \rightarrow \text{BEL}(k, \neg win)) \tag{9.14}$$

Formulas (9.1–9.7) describe the settings of the game. The initial setting is as described by Formula (9.1). During the game, tiles are placed in exactly one of the columns $1 \ldots m$ (Formulas (9.2–9.3)) and in every column exactly one tile type was previously placed (Formulas (9.4–9.5)). The boundary tiles are placed in columns 0 and $m + 1$ (Formula (9.6)) and nothing changes in columns where no tile is placed during the game (Formula (9.7)).

Formulas (9.8–9.11) describe the rules of the game. Tiles are placed from bottom to top, row by row from left to right (Formula (9.8)); thus, the first conjunct of Formula (9.8) represents the flipping of one row to the next. The players alternate (Formula (9.9)). Tiles that are placed have to match adjacent tiles (Formulas (9.10–9.11)). Formula (9.12) ensures that all possible moves by player B are encoded in the model.

Formula (9.13) gives properties of states that can be marked as winning positions for player A and Formula (9.14) conveys that all states reached after $\geq N$ rounds cannot be winning positions for A. Similarly to Blackburn *et al.* (2002, Lemma 6.1), one can force exponentially deep models of TEAMLOG for satisfying some specific formulas of depth ≤ 2. Specifically, to enumerate the states according to rounds of the game we will need the following additional formula:

$$\bigwedge_{j=1}^{N} \neg q_j \wedge \text{C-BEL}'_{\{1,2\}}\left(INC_0 \wedge \bigwedge_{j=1}^{N-1} INC_1(j)\right) \tag{9.15}$$

where:

$$INC_0 \equiv \neg q_1 \rightarrow \left(\text{BEL}(1, q_1) \wedge\right.$$

$$\bigwedge_{j=2}^{N}((q_j \rightarrow \text{BEL}(1, q_j)) \wedge (\neg q_j \rightarrow \text{BEL}(1, \neg q_j)))\right) \tag{9.16}$$

$$INC_1(i) \equiv \left(\neg q_{i+1} \wedge \bigwedge_{j=1}^{i} q_j\right) \rightarrow \text{BEL}\left(2, q_{i+1} \wedge \bigwedge_{j=1}^{i} \neg q_j \wedge\right.$$

$$\bigwedge_{j=i+2}^{N}((q_j \rightarrow \text{BEL}(2, q_j)) \wedge (\neg q_j \rightarrow \text{BEL}(2, \neg q_j)))\right) \tag{9.17}$$

Formula (9.15) enforces that the root of the model receives a number $(0 \ldots 0)_2$ and worlds corresponding to states in subsequent rounds of the game receive subsequent numbers in binary representation. The formula INC_0 is responsible for increasing even numbers and $INC_1(i)$ is responsible for increasing odd numbers ending with a sequence of i copies of 1 and having 0 at the position $i + 1$.

The formula $\varphi(\mathcal{G})$ is the conjunction of Formulas (9.1–9.15) and it is of size polynomial with respect to m. It can be easily seen that if A has a winning strategy in a particular game, the formula $\varphi(\mathcal{G})$ is satisfiable in a model built on the basis of a game tree for this game. Edges corresponding to turns of player A are the basis for accessibility relation B_1 and those corresponding to turns of player B are the basis for accessibility relation B_2. To satisfy the properties of the model, B_1 and B_2 are extended by identity in worlds that violate the seriality property. All other relations B_i and I_i are set to identity and relations G_i are set to $\varnothing$. The valuation of propositional variables in the worlds of the model is automatically determined by the description of the situation in the corresponding states of the game.

On the other hand, if $\varphi(\mathcal{G})$ is satisfiable, A can use a model of $\varphi(\mathcal{G})$ as a guide for his winning strategy. At the beginning, he/she chooses a transition (represented by accessibility relation B_1) to a world where win is true, and plays accordingly. Player A does analogously in all subsequent rounds of the game. He/she can track the worlds corresponding to states of subsequent rounds of the game, by following relations B_1 and B_2 alternatingly. Notice for all worlds v corresponding to states where A is to play and where A has a winning strategy (that is win is true) it must be $(v, v) \notin B_1$, as guaranteed by Formula (9.9). The same holds for B_2 and states where B is to play. Notice also that Formula (9.14) guarantees that A will reach a winning position in a finite number of steps if he/she plays as described above.

9.4.2 Effect of Bounding the Number of Propositional Atoms for TEAMLOG

If the number of propositional atoms is bounded by 1, the complexity of the satisfiability problem for TEAMLOG remains EXPTIME-hard. This can be easily shown by using an analogous technique to that described in Halpern (1995). The idea is to substitute propositional symbols used in the proof of Theorem 9.5, by the so-called pp-like formulas, that would have similar properties as propositional atoms (in terms of independence of their valuations in the worlds of a model). Suppose that propositional atoms are denoted by q_j. Then a pp-like formula replacing the propositional symbol q_j is $\neg OP(k, \neg p \wedge \neg BEL^j(1, \neg p))$, where $OP(k, \cdot)$ is any modal operator not used in the proof of Theorem 9.5.[10] See Halpern (1995) for additional details and an extended discussion of using pp-like formulas.

---

[10] Note that this argument will not work for the logic KD_1 with group operator M-INT, nor will it work for the logic $KD45_2$ with group operator C-BEL, because there is no 'free' modal operator left to be used as $OP(k, \cdot)$ for these cases. We do not know yet what would be the complexity of the satisfiability problem for these logics when the number of propositional atoms is bounded.

Similarly to the case of TEAMLOG^{ind}, we can show that if bounding the number of propositional atoms is combined with bounding the modal depth of formulas, the complexity is reduced to linear time. The proof is analogous to the one for TEAMLOG^{ind}.

Theorem 9.6 *For any fixed $k, l \geq 1$, if the number of propositional atoms is bounded by l and the modal depth of formulas is bounded by k, then the satisfiability problem for* TEAMLOG *can be solved in linear time.*

9.4.3 Effect of Restricting the Modal Context for TEAMLOG

In the previous sections we concluded that reducing the modal depth of formulas to 2 cannot help in reducing the complexity below EXPTIME, and even reducing the number of propositional atoms to 1 does not suffice. Only by combining the two types of restrictions did we get anywhere, namely to linear time, but even then we are left with a constant that depends exponentially on the number of propositions and the modal depth, so that using this restriction may not be tractable in practice. To solve this problem, Dziubiński (2007) presents a new approach to restricting the language of multi-modal logics with iterated modalities such as common belief and mutual intention. His restriction can be seen as generalization of restricting the modal depth of the formulas. It leads to a reduction of the satisfiability problem of TEAMLOG from EXPTIME-complete (full language) to PSPACE-complete for formulas with what he calls *restricted modal context*. Moreover, this restriction, when combined with restricting the modal depth of formulas, leads to NPTIME-completeness of the satisfiability problem. The restrictions that Dziubiński proposes constrain the modal context of subformulas of formulas. For the language constrained this way, a properly extended tableau method can be used to decide the satisfiability of the formulas.

Let us first define the notion of modal context and associated notions, for a general language of multi-modal logic. First we need a notion of *modal context of a formula within a formula*. Let $\mathcal{L}[\Omega]$ be a multi-modal language based on a set of unary modal operators Ω, and let Ω^* denote the set of all finite sequences over Ω.

Definition 9.9 (Modal context of a formula within a formula) *Let $\{\varphi, \xi\} \subseteq \mathcal{L}[\Omega]$. The* modal context of formula ξ within formula φ, *denoted as* $\text{cont}\,(\xi, \varphi) \subseteq \Omega^*$, *is defined inductively as follows*:

- $\text{cont}\,(\xi, \varphi) = \varnothing$, if $\xi \notin \text{Sub}(\varphi)$;
- $\text{cont}\,(\varphi, \varphi) = \{\varepsilon\}$;
- $\text{cont}\,(\xi, \neg\psi) = \text{cont}\,(\xi, \psi)$, if $\xi \neq \neg\psi$;
- $\text{cont}\,(\xi, \psi_1 \wedge \psi_2) = \text{cont}\,(\xi, \psi_1) \cup \text{cont}\,(\xi, \psi_2)$, if $\xi \neq \psi_1 \wedge \psi_2$;
- $\text{cont}\,(\xi, \Box\psi) = \Box \cdot \text{cont}\,\big(\xi, \psi_j\big)$, if $\xi \neq \Box\psi$ and $\Box \in \Omega$, where $\Box \cdot S = \{\Box \cdot s : s \in S\}$, for $\Box \in \Omega$ and $S \subseteq \Omega^*$.

Now we can introduce the notion of modal context restrictions as in Dziubiński (2007).

Definition 9.10 (Modal context restriction) A modal context restriction *is a set of sequences over Ω, $S \subseteq \Omega^*$, constraining possible modal contexts of subformulas within formulas.*

We say that a formula $\varphi \in \mathcal{L}[\Omega]$ satisfies a modal context restriction $S \subseteq \Omega^*$ iff for all $\xi \in \mathrm{Sub}(\varphi)$ it holds that $\mathrm{cont}\,(\xi, \varphi) \subseteq S$.

Dziubiński (2007) studied two modal context restrictions for TEAMLOG, namely $\mathbf{R}_1$ and $\mathbf{R}_2$ defined below. For simplicity, we assume that formulas of the form C-BEL$_{\{j\}}(\psi)$ are replaced by BEL(j, ψ) and do not occur in the language; hence for formulas C-BEL$_G(\psi)$, it is always the case that $|G| \geq 2$.

Definition 9.11 (Restriction $\mathbf{R}_1$) *Let:*

$$\mathbf{R}_1 = \tau^* \setminus \left(\tau^* \cdot \left[\bigcup_{G \in P(\mathcal{A}) \setminus \{\varnothing\}} (S_{\mathrm{I}}(G) \cup S_{\mathrm{IB}}(G)) \cup \bigcup_{G \in P(\mathcal{A}), |G| \geq 2} S_{\mathrm{B}}(G) \right] \cdot \tau^* \right)$$

where:

$$S_{\mathrm{IB}}(G) = \bigcup_{\substack{H \in P(\mathcal{A}) \\ H \cap G \neq \varnothing}} \text{M-INT}_G \cdot \text{C-BEL}_H \cdot T_{\mathrm{I}}(G \cap H) \cup \bigcup_{j \in G} \text{M-INT}_G \cdot \text{BEL}_j \cdot T_{\mathrm{I}}(\{j\}) \quad \text{and}$$

$S_{\mathrm{B}}(G) = \text{C-BEL}_G \cdot T_{\mathrm{B}}(G);$

$T_{\mathrm{B}}(G) = \{\text{BEL}_j : j \in G\} \cup \{\text{C-BEL}_H : H \in P(\mathcal{A}), H \cap G \neq \varnothing\};$

$S_{\mathrm{I}}(G) = \text{M-INT}_G \cdot T_{\mathrm{I}}(G);$

$T_{\mathrm{I}}(G) = \{\text{INT}_j : j \in G\} \cup \{\text{M-INT}_H : H \in P(\mathcal{A}), H \cap G \neq \varnothing\}.$

The set of formulas in $\mathcal{L}$ satisfying restriction $\mathbf{R}_1$ will be denoted by $\mathcal{L}_{\mathbf{R}_1}$.

The sets S_B, S_I, and S_{IB} describe sequences of modal operators in the context of subformulas within formulas that are forbidden by restrictions $\mathbf{R}_1$. Let us explain the definition by looking at some types of forbidden formulas, one by one, and provide examples of each in turn. In order to avoid one source of complexity, intuitively speaking, modal context restriction $\mathbf{R}_1$ forbids formulas of the form $\text{INT}(j, \xi)$ and M-INT$_H(\xi)$ with $j \in G$ and $H \cap G \neq \varnothing$ with a direct context 'around' it of operator M-INT$_G(\cdot)$. Therefore, the following formulas do not satisfy the restriction $\mathbf{R}_1$:

$$\text{M-INT}_{\{1,2\}}(\text{INT}(1, p))$$

$$\text{M-INT}_{\{1,2\}}(q \vee \text{INT}(2, p))$$

$$\text{M-INT}_{\{1,2\}}(\text{M-INT}_{\{2,3,4\}}(p))$$

$$\text{M-INT}_{\{1,2\}}(\text{M-INT}_{\{3,4\}}(q \wedge \text{INT}(3, r)))$$

Similarly, formulas of the form BEL(j, ξ) and C-BEL$_H(\xi)$ with $j \in G$ and $H \cap G \neq \varnothing$ with a direct context of operator C-BEL$_G(\cdot)$ are forbidden by $\mathbf{R}_1$. So, for example, the following formulas do not satisfy the restriction $\mathbf{R}_1$:

$$\text{C-BEL}_{\{1,2\}}(\text{BEL}(1, p))$$

$$\text{C-BEL}_{\{1,2\}}(q \vee \text{BEL}(2, p))$$

$$C\text{-BEL}_{\{1,2\}}(C\text{-BEL}_{\{2,3,4\}}(p))$$

$$C\text{-BEL}_{\{1,2\}}(C\text{-BEL}_{\{3,4\}}(q \wedge \text{BEL}(3,r)))$$

Additionally, the following formulas are forbidden by restriction $\mathbf{R}_1$ due to the S_{IB}-sequences of M-INT governing BEL, governing another M-INT or INT in them, where groups overlap:

$$M\text{-INT}_{\{1,2\}}(\text{BEL}(1, \text{INT}(1, p)))$$

$$M\text{-INT}_{\{1,2\}}(\text{BEL}(2, q \vee \text{INT}(2, p)))$$

$$M\text{-INT}_{\{1,2\}}(\text{BEL}(2, M\text{-INT}_{\{2,3,4\}}(p)))$$

This restriction is needed due to the fact that TEAM-LOG $\vdash \text{INT}(j, \psi) \leftrightarrow \text{BEL}$ $(j, \text{INT}(j, \psi))$. In contrast, the following formulas are allowed by $\mathbf{R}_1$, so they are in language $\mathcal{L}_{\mathbf{R}_1}$:

$$M\text{-INT}_{\{1,2\}}(\text{BEL}(1, \text{INT}(2, p)))$$

$$C\text{-BEL}_{\{1,2\}}(\text{BEL}(3, C\text{-BEL}_{\{1,2\}}(p)))$$

$$M\text{-INT}_{\{1,2\}}(q) \wedge C\text{-BEL}_{\{1,2\}}(M\text{-INT}_{\{1,2\}}(q))$$

For further reductions in the complexity of TEAMLOG-satisfiability, let us define an even graver restriction on the modal context of formulas.

Definition 9.12 (Restriction $\mathbf{R}_2$) *Let:*

$$\mathbf{R}_2 = \tau^* \setminus \left(\tau^* \cdot \left[\bigcup_{G \in \mathsf{P}(\mathcal{A}) \setminus \{\varnothing\}} (S_{\text{I}}(G) \cup S_{\text{IB}}(G)) \cup \bigcup_{G \in \mathsf{P}(\mathcal{A}), |G| \geq 2} \tilde{S}_{\text{B}}(G) \right] \cdot \tau^* \right)$$

where:

$$\tilde{S}_{\text{B}}(G) = C\text{-BEL}_G \cdot \left(\{\text{GOAL}_j : j \in G\} \cup \bigcup_{O \in \{\text{B,I}\}} T_O(G) \right)$$

Here, S_{IB}, S_{I}, T_B and T_I are defined like in the case of restriction $\mathbf{R}_1$ in Definition 9.11. The set of formulas in $\mathcal{L}$ satisfying restriction $\mathbf{R}_2$ will be denoted by $\mathcal{L}_{\mathbf{R}_2}$.

Similarly as for $\mathbf{R}_1$ but even more restrictively, the sets S_B, S_I, S_{IB} and $\tilde{S}_B$ describe sequences of modal operators in the context of subformulas within formulas that are forbidden by restriction $\mathbf{R}_2$.

Modal context restriction $\mathbf{R}_2$ is a refinement of $\mathbf{R}_1$: it is easy to see that $\mathcal{L}_{\mathbf{R}_2} \subseteq \mathcal{L}_{\mathbf{R}_1}$, while restriction $\mathbf{R}_2$ additionally forbids formulas of the form $\text{INT}(j, \psi)$, $\text{GOAL}(j, \psi)$, and $M\text{-INT}_H(\psi)$ within the direct context of operator $C\text{-BEL}_G(\cdot)$, in cases where $j \in G$ and $H \cap G \neq \varnothing$. Thus, in particular, the following formula, which satisfies $\mathbf{R}_1$, still violates $\mathbf{R}_2$:

$$M\text{-INT}_{\{1,2\}}(q) \wedge C\text{-BEL}_{\{1,2\}}(M\text{-INT}_{\{1,2\}}(q))$$

The reader will recognize the right-hand side of axiom **M3** from Chapter 3 here: the definition of collective intention. This means that $\mathbf{R}_1$ does not rule out collective intention, while $\mathbf{R}_2$ does. Now let us investigate the effects that both restrictions have on the complexity of TEAMLOG satisfiability, beginning with the most restricted language. Proofs are skipped here, but can be found in Dziubiński (in preparation). The complexity results for $\mathbf{R}_2$ are:

- Checking TEAMLOG satisfiability of formulas from $\mathcal{L}_{\mathbf{R}_2}$ is PSPACE-complete.
- Checking TEAMLOG satisfiability of formulas from $\mathcal{L}_{\mathbf{R}_2}$ with modal depth bounded by a constant k is NPTIME-complete.
  Moreover, the complexity is $\mathcal{O}(((2|\mathcal{A}|+1)|\varphi|)^{k^{(|\mathcal{A}|+1)}})$, where $|\varphi|$ is the size of the input formula.

In the case of $\mathcal{L}_{\mathbf{R}_1}$ it is possible to construct a formula with modal depth 2 which enforces an exponentially long path within its interpretation. Nevertheless the complexity results for $\mathbf{R}_1$ are:

- Checking TEAM-LOG satisfiability of formulas from $\mathcal{L}_{\mathbf{R}_1}$ is PSPACE-complete and it remains so even if modal depth of formulas is bounded by a constant $k \geq 2$.

Although, as we mentioned earlier, it is possible to enforce exponentially deep models using formulas from $\mathcal{L}_{\mathbf{R}_1}$ it is still possible to solve the satisfiability problem using polynomial space. The idea is to combine the tableau method with Savitch's $\mathcal{O}(\log^2(n))$ algorithm for reachability on directed graphs (Savitch, 1970), cf. Dziubiński (in preparation) for details).[11]

Lastly, modal context restriction $\mathbf{R}_1$ can be refined, so that additionally bounding modal depth of formulas leads to NPTIME-completeness of the satisfiability problem. In this way, Dziubiński (2007) obtains an NPTIME satisfiability problem without ruling out collective intention (as $\mathbf{R}_2$ does). Given a formula φ, let $PT(\varphi)$ denote the set of subformulas of φ taken with respect to propositional operators only. The refinement is as follows. Whenever we have a formula φ containing a subformula of the form C-BEL$_G(\psi)$ that violates modal context restriction $\mathbf{R}_2$ (that is $PT(\psi)$ contains a formula of the form INT(j, ξ), GOAL(j, ξ) or M-INT$_H(\xi)$ with $j \in G$ or $H \cap G \neq \varnothing$), then the set of such formulas (for each $j \in G$) is bounded by a constant; here, M-INT$_H(\xi)$ counts for each $j \in H$. For example, we could allow for only one such formula which is sufficient to allow for collective intention.

9.5 Discussion and Conclusions

This chapter deals with the complexity of two important components of TEAMLOG. The first one, TEAMLOG^{ind}, covering agents' individual attitudes and their interdependencies,

---

[11] This result is surprising for those of us who remember that in the case of PDL, which has an EXPTIME-complete satisfiability problem, 'exponentially deep branches in the tableau spell trouble' (Blackburn *et al.*, 2002); still we shouldn't get our hopes up that satisfiability for PDL could be in PSPACE as well, for that would imply PSPACE = EXPTIME, an unlikely answer to a longstanding open problem.

was proved to be PSPACE-complete. The second one, TEAMLOG itself, dealing with the team attitude par excellence, collective intention, turns out to be EXPTIME-complete.

Importantly, however, our results have a more general impact. The tableau methods can be adapted to the non-temporal parts of other multi-modal logics similar in spirit to ours, such as the KARO framework (Aldewereld *et al.*, 2004). Note that the PSPACE-completeness of TEAMLOG^{ind} is not a disadvantage of our particular theory. The standard BDI logics for individual attitudes based on a normal modal logic, even without interdependency axioms, have a similar complexity.

Decidability of TEAMLOG already follows from the completeness proof in Chapter 3. More precisely, it is EXPTIME-complete. Again, this high complexity is not a quirk of our particular choice for TEAMLOG. For example, alternating-time temporal logic (ATL), used for specifying coalitional power, has an EXPTME-complete satisfiability problem, whether the group is fixed or not (Drimmelen, 2003; Walther *et al.*, 2006). Goranko and Shkatov have recently proved that a multi-agent logic similar to TEAMLOG (that is epistemic logic including common knowledge and distributed knowledge for different groups) remains EXPTIME-complete even when combined with branching time temporal logic (Goranko and Shkatov, 2009). In such endeavors, further research will include applications of general techniques to combine modal logics, such as *fibering* (Gabbay, 1998).

How to make these complex teamwork logics more manageable? As with other modal logics, an option would be to develop a variety of different algorithms and heuristics, each performing well on a limited class of inputs. For example, it is known that restricting the number of propositional atoms and/or the depth of modal nesting may reduce the complexity (Halpern, 1995; Hustadt and Schmidt, 1997). We explored these possibilities in this chapter for both for TEAMLOG^{ind} and TEAMLOG. Also, when considering specific applications, it is possible to reduce some of the infinitary character of collective beliefs and intentions to more manageable proportions (Fagin *et al.*, 1995, Chapter 11). For artificial intelligence applications it is particularly interesting to restrict the language to Horn-like formulas (Nguyen, 2005). Such restrictions are essential in the context of collective commitment, in order to obtain system specifications of lower complexity. See Chapters 4 and 7 for some possible other avenues to reduce complexity by domain-specific simplifications.

Another technique conclusive in reducing the complexity could depend on simplifying multi-modal theories of collective attitudes using approximations in the spirit of rough set theory introduced by Pawlak (1981, 1991). Pawlak's influential ideas, developed over the last 25 years by many researchers, appeared very useful, among others, in the context of reducing the complexity of reasoning over large data sets.

It seems rather natural to extend his approach to multi-modal logics. In fact, logical approximations have been considered in various papers (Cadoli, 1995, Doherty *et al.* 2001, Kautz and Selman, 1996 and Lin, 2000) and in a book (Doherty *et al.*, 2006). It can be shown that the approximations considered in Doherty *et al.*, (2001) and Lin, (2000) are as strong as the rough approximations introduced by Pawlak. Approximate reasoning has also been fruitfully applied to theories like TEAMLOG in Doherty *et al.* (2007, 2003), Dunin-Kęplicz, Nguyen and Szałas (2010, 2009a) and Dunin-Kęplicz and Szałas (2007, 2010). Complexity studies about an approximate analogue of TEAMLOG^{ind} are underway.

Appendix A

A.1 Axiom Systems

For ease of reference, we recall the system TEAMLOG^{ind} for individual attitudes and their interdependencies, followed by our additional axioms and rules for group attitudes. These axioms and rules, together forming TEAMLOG, are explained in Chapters 2 and 3. All axiom systems introduced here are based on the finite set $\mathcal{A}$ of n agents.

A.1.1 Axioms for Individual and Collective Attitudes

General Axiom and Rule

The following axiom and rule, covering propositional reasoning, form part and parcel of any system of normal modal logic:

P1   All instances of propositional tautologies
PR1   From φ and $\varphi \rightarrow \psi$, derive ψ (Modus Ponens)

Axioms and Rules for Individual Belief

The well-known system KD45$_n$ consists of the following for each $i \in \mathcal{A}$:

A2 $\text{BEL}(i, \varphi) \wedge \text{BEL}(i, \varphi \rightarrow \psi) \rightarrow \text{BEL}(i, \psi)$ (Belief Distribution)
A4 $\text{BEL}(i, \varphi) \rightarrow \text{BEL}(i, \text{BEL}(i, \varphi))$ (Positive Introspection)
A5 $\neg\text{BEL}(i, \varphi) \rightarrow \text{BEL}(i, \neg\text{BEL}(i, \varphi))$ (Negative Introspection)
A6 $\neg\text{BEL}(i, \bot)$ (Consistency)
R2   From φ infer $\text{BEL}(i, \varphi)$ (Belief Generalization)

Teamwork in Multi-Agent Systems: A Formal Approach   Barbara Dunin-Kęplicz and Rineke Verbrugge
© 2010 John Wiley & Sons, Ltd

Axioms for Individual Motivational Operators

For goals, we take the system K_n and for intentions the system KD_n, as follows, for each $i \in \mathcal{A}$:

$A2_G$	$GOAL(i, \varphi) \wedge GOAL(i, \varphi \to \psi) \to GOAL(i, \psi)$	(Goal Distribution)
$A2_I$	$INT(i, \varphi) \wedge INT(i, \varphi \to \psi) \to INT(i, \psi)$	(Intention Distribution)
$R2_G$	From φ infer $GOAL(i, \varphi)$	(Goal Generalization)
$R2_I$	From φ infer $INT(i, \varphi)$	(Intention Generalization)
$A6_I$	$\neg INT(i, \bot)$ for $i = 1, \ldots, n$	(Intention Consistency)

Interdependencies Between Intentions and Other Attitudes

For each $i \in \mathcal{A}$:

$A7_{GB}$	$GOAL(i, \varphi) \to BEL(i, GOAL(i, \varphi))$	(Positive Introspection for Goals)
$A7_{IB}$	$INT(i, \varphi) \to BEL(i, INT(i, \varphi))$	(Positive Introspection for Intentions)
$A8_{GB}$	$\neg GOAL(i, \varphi) \to BEL(i, \neg GOAL(i, \varphi))$	(Negative Introspection for Goals)
$A8_{IB}$	$\neg INT(i, \varphi) \to BEL(i, \neg INT(i, \varphi))$	(Negative Introspection for Intentions)
$A9_{IG}$	$INT(i, \varphi) \to GOAL(i, \varphi)$	(Intention implies Goal)

By TEAMLOG^{ind} we denote the axiom system consisting of all the above axioms and rules for individual beliefs, goals and intentions as well as their interdependencies.

Axioms and Rule For General ('Everyone') and Common Belief

C1	$E\text{-}BEL_G(\varphi) \leftrightarrow \bigwedge_{i \in G} BEL(i, \varphi)$	(General Belief)
C2	$C\text{-}BEL_G(\varphi) \leftrightarrow E\text{-}BEL_G(\varphi \wedge C\text{-}BEL_G(\varphi))$	(Common Belief)
RC1	From $\varphi \to E\text{-}BEL_G(\psi \wedge \varphi)$ infer $\varphi \to C\text{-}BEL_G(\psi)$	(Induction Rule)

Axioms and Rule for General, Mutual and Collective Intentions

M1	$E\text{-}INT_G(\varphi) \leftrightarrow \bigwedge_{i \in G} INT(i, \varphi)$	(General Intention)
M2	$M\text{-}INT_G(\varphi) \leftrightarrow E\text{-}INT_G(\varphi \wedge M\text{-}INT_G(\varphi))$	(Mutual Intention)
M3	$C\text{-}INT_G(\varphi) \leftrightarrow M\text{-}INT_G(\varphi) \wedge C\text{-}BEL_G(M\text{-}INT_G(\varphi))$	(Collective Intention)
RM1	From $\varphi \to E\text{-}INT_G(\psi \wedge \varphi)$ infer $\varphi \to M\text{-}INT_G(\psi)$	(Induction Rule)

By TEAMLOG we denote the union of TEAMLOG^{ind} with the above axioms and rules for general and common beliefs and for general, mutual and collective intentions.

A.1.2   Axioms for Social Commitments

Here follows the defining axiom for social commitments with respect to propositions:

SC1

$$COMM(i, j, \varphi) \leftrightarrow INT(i, \varphi) \wedge GOAL(j, done(i, \texttt{stit}(\varphi))) \wedge$$

$$C\text{-}BEL_{\{i, j\}}(INT(i, \varphi) \wedge GOAL(j, done(i, \texttt{stit}(\varphi))))$$

where $done(i, \texttt{stit}(\varphi))$ means that agent i has just seen to it that φ was achieved.

Social commitments with respect to actions are defined by the axiom:

SC2

$$\text{COMM}(i, j, \alpha) \leftrightarrow \text{INT}(i, \alpha) \wedge \text{GOAL}(j, done(i, \alpha)) \wedge$$

$$\text{C-BEL}_{\{i,j\}}(\text{INT}(i, \alpha) \wedge \text{GOAL}(j, done(i, \alpha)))$$

where $done(i, \alpha)$ means that agent i has just executed action α.

A.1.3 Tuning Schemes for Social and Collective Attitudes

Collective intention

$\mathbf{M3}^{\text{schema}}$ $\text{C-INT}_G(\varphi) \leftrightarrow \text{M-INT}_G(\varphi) \wedge awareness_G(\text{M-INT}_G(\varphi))$

Social commitment

$$\text{COMM}(i, j, \alpha) \leftrightarrow \text{INT}(i, \alpha) \wedge \text{GOAL}(j, done(i, \alpha)) \wedge$$

$$awareness_{\{i,j\}}(\text{INT}(i, \alpha) \wedge \text{GOAL}(j, done(i, \alpha)))$$

Collective commitment

$\text{C-COMM}_{G,P}(\varphi) \leftrightarrow$

$\text{M-INT}_G(\varphi) \wedge \{awareness_G^1(\text{M-INT}_G(\varphi))\} \wedge$

$constitute(\varphi, P) \wedge \{awareness_G^2(constitute(\varphi, P))\} \wedge$

$\bigwedge\limits_{\alpha \in P} \bigvee\limits_{i,j \in G} \text{COMM}(i, j, \alpha) \wedge$

$\{awareness_G^3(\bigwedge\limits_{\alpha \in P} \bigvee\limits_{i,j \in G} \text{COMM}(i, j, \alpha)) / \bigwedge\limits_{\alpha \in P} \bigvee\limits_{i,j \in G} awareness_G^3(\text{COMM}(i, j, \alpha))\}$

A.1.4 Axioms for Exemplary Collective Commitments

Robust collective commitment $\text{R-COMM}_{G,P}$

1. Collective intention within the team
2. Correct plan P leading to φ
3. Collective awareness of correctness of P
4. Social commitments for all actions in P
5. Detailed collective awareness about social commitments

$$\text{R-COMM}_{G,P}(\varphi) \leftrightarrow \text{C-INT}_G(\varphi) \wedge$$

$$constitute(\varphi, P) \wedge$$

$$\text{C-BEL}_G(constitute(\varphi, P)) \wedge$$

$$\bigwedge_{\alpha \in P} \bigvee_{i,j \in G} \text{COMM}(i, j, \alpha) \wedge$$

$$\bigwedge_{\alpha \in P} \bigvee_{i,j \in G} \text{C-BEL}_G(\text{COMM}(i, j, \alpha))$$

Strong collective commitment $\text{S-COMM}_{G,P}$

1. Collective intention within the team
2. Correct plan P leading to φ
3. Collective awareness of correctness of P
4. Social commitments for all actions in P
5. Global collective awareness about existence of social commitments

$$\text{S-COMM}_{G,P}(\varphi) \leftrightarrow \text{C-INT}_G(\varphi) \wedge$$

$$constitute(\varphi, P) \wedge$$

$$\text{C-BEL}_G(constitute(\varphi, P)) \wedge$$

$$\bigwedge_{\alpha \in P} \bigvee_{i,j \in G} \text{COMM}(i, j, \alpha) \wedge$$

$$\text{C-BEL}_G(\bigwedge_{\alpha \in P} \bigvee_{i,j \in G} \text{COMM}(i, j, \alpha))$$

Weak collective commitment $\text{W-COMM}_{G,P}$

1. Collective intention within the team
2. Correct plan P leading to φ
3. Social commitments for all actions in P
4. Global collective awareness about existence of social commitments

$$\text{W-COMM}_{G,P}(\varphi) \leftrightarrow \text{C-INT}_G(\varphi) \wedge$$

$$constitute(\varphi, P) \wedge$$

$$\bigwedge_{\alpha \in P} \bigvee_{i,j \in G} \text{COMM}(i, j, \alpha) \wedge$$

$$\text{C-BEL}_G(\bigwedge_{\alpha \in P} \bigvee_{i,j \in G} \text{COMM}(i, j, \alpha))$$

Team commitment $\text{T-COMM}_{G,P}(\varphi)$

1. Collective intention within the team
2. Correct plan P leading to φ

3. Social commitments for all actions in P

$$\text{T-COMM}_{G,P}(\varphi) \leftrightarrow \text{C-INT}_G(\varphi) \wedge$$

$$constitute(\varphi, P) \wedge$$

$$\bigwedge_{\alpha \in P} \bigvee_{i,j \in G} \text{COMM}(i, j, \alpha)$$

Distributed commitment $\quad \text{D-COMM}_{G,P}(\varphi)$

1. Correct plan P leading to φ
2. Social commitments for all actions in P

$$\text{D-COMM}_{G,P}(\varphi) \leftrightarrow constitute(\varphi, P) \wedge$$

$$\bigwedge_{\alpha \in P} \bigvee_{i,j \in G} \text{COMM}(i, j, \alpha).)$$

By TEAMLOG^{com} we denote the union of TEAMLOG with the above axioms for social and collective commitments. axioms for social commitments and chosen axioms for collective commitments (see Chapter 4).

A.1.5 Axioms and Rules for Dynamic Logic

P2 $\quad [do(i, \alpha)](\varphi \rightarrow \psi) \rightarrow ([do(i, \alpha)]\varphi \rightarrow [do(i, \alpha)]\psi)$ (Dynamic Distribution)
P3 $\quad [do(i, \texttt{confirm}(\varphi))]\psi \leftrightarrow (\varphi \rightarrow \psi)$
P4 $\quad [do(i, \alpha_1; \alpha_2)]\varphi \leftrightarrow [do(i, \alpha_1)][do(i, \alpha_2)]\varphi$
P5 $\quad [do(i, \alpha_1 \cup \alpha_2)]\varphi \leftrightarrow ([do(i, \alpha_1)]\varphi \wedge [do(i, \alpha_2)]\varphi$
P6 $\quad [do(i, \alpha^*)]\varphi \rightarrow \varphi \wedge [do(i, \alpha)][do(i, \alpha^*)]\varphi$ (Mix)
P7 $\quad (\varphi \wedge [do(i, \alpha^*)](\varphi \rightarrow [do(i, \alpha)]\varphi)) \rightarrow [do(i, \alpha^*)](\varphi)$ (Induction)
PR2 From φ, derive $[do(i, \alpha)]\varphi$ (Dynamic Necessitation)

By TEAMLOG^{dyn} we denote the union of TEAMLOG^{com} with the above axioms for dynamic operators (see Chapter 6).

Thus in general, we have TEAMLOG$^{ind} \subseteq$ TEAMLOG $\subseteq$ TEAMLOG$^{com} \subseteq$ TEAMLOG^{dyn}.

A.2 An Alternative Logical Framework for Dynamics of Teamwork: Computation Tree Logic

We chose to use dynamic logic in our teamwork theory TEAMLOG^{dyn}. Many BDI architectures are based on temporal logic: linear time was the model selected by Cohen and Levesque (1990), while Rao and Georgeff (1991) chose branching time. Lately, Alternating-Time Temporal Logics (ATL) have become popular in the literature (Jamroga and van der Hoek, 2004).

Below we present the temporal part of Rao and Georgeff's theory for readers who wish to adapt our teamwork theory to a temporal underlying semantics. For example, the definitions of collective commitments in terms of more basic attitudes, as presented in Chapter 4, may be combined with either choice, depending on the application.

As a reminder of Rao and Georgeff (1991), their temporal structure is a discrete tree branching towards the future, as in Computation Tree Logic (CTL), which is used for studying concurrent programs (see Emerson (1990) for a semantic and axiomatic treatment). The different branches in such a time tree denote the optional courses of events that can be chosen by an agent. An agent can perform primitive events that determine a next time point on a branch in the tree. The branch between a point and the next point is labeled with the primitive event leading to that point. For example, if there are two branches emanating from a single time point, one labeled 'go to dentist' and the other 'go shopping', then the agent has a choice of executing either of these events and moving to the next point along the associated branch. The temporal operators include $\mathbf{A}$ (φ) (in all paths through the point of reference φ holds), $\mathbf{E}$ $(\varphi) \equiv \neg\mathbf{A}$ $(\neg\varphi)$, $\Diamond\varphi$ (somewhere later on the same path, φ holds) and φ $\mathbf{U}$ ψ (φ *until* ψ, that is either φ holds forever on this path, or, as soon as it stops holding, ψ will hold). Formulas are divided into state formulas (which are true in a particular state) and path formulas (which are true along a certain path). Here follows our definition, which adapts Rao and Georgeff's single-agent definition to the n-agent case with a set of agents $\mathcal{A}$:

S1 each atomic proposition is a state formula
S2 if φ and ψ are state formulas, then so are $\neg\varphi$ and $\varphi \wedge \psi$
S3 if φ is a path formula then $\mathbf{A}$ (φ) and $\mathbf{E}$ (φ) are state formulas
S4 if φ is a state formula, then so are BEL(a, φ), GOAL(a, φ) and INT(a, φ) for all $a \in A$
P0 if φ and ψ are state formulas, then so are $\Diamond\varphi$ and φ $\mathbf{U}$ ψ

As to Kripke semantics, we consider each possible world to be a temporal tree structure as described above with a single past and branching-time future. Evaluation of formulas is with respect to a world w and a state s, using ternary accessibility relations B_i, D_i and I_i corresponding to agents' beliefs, goals (or desires) and intentions, all of which lead from a pair of a world and a state in it to a world. Evaluation of formulas at world-state pairs is defined in the obvious manner inspired by CTL and epistemic logic. Here we give only our n-agent adaptation of the definitions for beliefs, goals and intentions, where the expression $M, w_s \vDash \varphi$ is read as 'formula φ is satisfied by world w and state s in structure M'. For $i = 1, \ldots, n$ we have:

$$M, w_s \vDash \text{GOAL}(i, \varphi) \text{ iff } \forall v \text{ with } (w, s, v) \in D_i, \ M, v_s \vDash \varphi$$

$$M, w_s \vDash \text{INT}(i, \varphi) \text{ iff } \forall v \text{ with } (w, s, v) \in I_i, M, v_s \vDash \varphi$$

The full definition of formula evaluation can be found in Rao and Georgeff (1995b) and some examples are given in Rao and Georgeff (1991). We will need this notion of possible worlds below for describing commitment strategies.

Rao and Georgeff (1995b) give an axiomatization of a basic BDI-logic for the single-agent case, which includes all CTL-axioms for the temporal component. For the epistemic operator BEL, the modal system KD45 for a single agent is used. For the motivational operators GOAL and INT, their axioms include the system KD. However, it was argued in Chapter 3 that an agent's goals are not necessarily consistent with one another. Rao and Georgeff prove soundness and completeness of their basic BDI-logic and some extensions with respect to suitable classes of models by a tableau method and also give decidability results using a small model theorem.

A.2.1 Commitment Strategies

In the main part of this book, we mention commitment strategies only briefly in Chapter 5. The key point is whether and in which circumstances an agent can drop a social commitment. If such a situation arises, the next question is how to deal with it responsibly. All three kinds of agents communicate with their partner after dropping a social commitment. We give the formal definitions below (see also Chapter 8 plus Dunin-Kęplicz and Verbrugge (1996, 1999) and Rao and Georgeff (1991)).

We define three kinds of agents according to the strength with which they maintain their social commitments. The definitions are inspired by those in Rao and Georgeff (1991) for intention strategies. The need for agents' responsible behavior led us to include additionally the social aspects of communication and coordination. We assume that the commitment strategies are an immanent property of the individual agent and that they do not depend on the goal to which the agent is committed, nor on the other agent to whom it is committed. We also assume that each agent is aware which commitment strategies are adopted by the agents in the group. This 'meta-knowledge' ensures proper re-planning and coordination (Dunin-Kęplicz and Verbrugge, 1996). Here follow some definitions based on the branching time framework introduced above.

The strongest commitment strategy is followed by the *blindly committed* agent, who maintains its commitments until it actually believes that they have been achieved. Formally:

$$\text{COMM}(a, b, \varphi) \rightarrow \mathbf{A} \, (\text{COMM}(a, b, \varphi) \, \mathbf{U} \, \text{BEL}(a, \varphi))$$

Single-minded agents may drop social commitments when they do not believe anymore that the commitment is realizable. However, as soon as the agent abandons a commitment, some communication and coordination with the other agent is needed:

$$\text{COMM}(a, b, \varphi) \rightarrow$$
$$\mathbf{A} \, [\text{COMM}(a, b, \varphi) \, \mathbf{U}$$
$$\{\text{BEL}(a, \varphi) \vee$$
$$(\neg \text{BEL}(a, \mathbf{E} \, \diamond \varphi) \wedge$$
$$\textit{done}(\texttt{communicate}(a, b, \neg \text{BEL}(a, \mathbf{E} \, \diamond \varphi))) \wedge$$
$$\textit{done}(\texttt{coordinate}(a, b, \varphi)))\}]$$

For *open-minded* agents, the situation is similar as for single-minded ones, except that they can also drop social commitments if they do not aim for the respective goal anymore. As in the case of single-minded agents, communication and coordination will be involved, as expressed by the axiom:

$$\text{COMM}(a, b, \varphi) \rightarrow$$

$$\mathbf{A} \,[\text{COMM}(a, b, \varphi) \,\mathbf{U}$$

$$\{\text{BEL}(a, \varphi) \lor$$

$$(\neg\text{BEL}(a, \mathbf{E} \diamond \varphi) \land$$

$$done(\texttt{communicate}(a, b, \neg\text{BEL}(a, \mathbf{E} \diamond \varphi))) \land$$

$$done(\texttt{coordinate}(a, b, \varphi)))$$

$$\lor(\neg\text{GOAL}(a, \mathbf{E} \diamond \varphi) \land$$

$$done(\texttt{communicate}(a, b, \neg\text{GOAL}(a, \mathbf{E} \diamond \varphi))) \land$$

$$done(\texttt{coordinate}(a, b, \varphi)))\}]$$

A.2.2 The Blocking Case Formalized in the Temporal Language

In Chapter 6, the most serious case one could meet during reconfiguration is the case where an objectively failed task blocks the system's goal φ (see Section 6.5.2.2). To formalize this case in a more subtle way than in Chapter 6 and to prove consequences, a more extended language is needed than the dynamic one used there. Here we give some hints as to how this may be done. The Kripke model is extended with a discrete temporal structure branching towards the future (as in CTL) and the language includes operators $\mathbf{E}$ (eventually) for 'in future on some branch through the present point' (with its dual $\mathbf{A}$ for all branches), $\mathbf{P}$ for 'somewhere in the past' and $\diamond$ for 'in future somewhere on the current branch' (with its dual $\Box$).

Thus, at a moment where action α has not succeeded before, j just failed executing it and no agent will ever achieve it we have:

$$\mathcal{M}, w_t \vDash \neg\exists i \mathbf{P} succ(i, \alpha) \land failed(j, \alpha) \land \neg\exists i \,(\mathbf{E} \diamond (succ(i, \alpha)))$$

We then define 'α is necessary for achieving φ' formally as

$$\mathcal{M}, w_t \vDash \neg\exists i \mathbf{P}(succ(i, \alpha)) \rightarrow \neg\varphi$$

It follows by temporal logic from both formulas that:

$$\mathcal{M}, w_t \vDash \mathbf{A}\Box\neg\varphi$$

that is φ will never hold. Thus, if it is discovered that a failed action blocks the overall goal in the above way, the system fails and neither a collective intention nor an evolved collective commitment towards it will be established.

Bibliography

Aaqvist, L. (1984). Deontic logic. In Gabbay, D. and Guenthner, F. (Eds), *Handbook of Philosophical Logic*, Volume III, pages 605–714. Reidel, Dordrecht, The Netherlands.

Ågotnes, T. and Alechina, N. (2007a). The dynamics of syntactic knowledge. *Journal of Logic and Computation*, **17**(1): 83–116.

Ågotnes, T. and Alechina, N. (2007b). Full and relative awareness: A decidable logic for reasoning about knowledge of unawareness. In *Proceedings of TARK XI*, pages 6–14. ACM Press, New York, NY, USA.

Ågotnes, T., van der Hoek, W. and Wooldridge, M. (2008). Quantifying over coalitions in epistemic logic. In Padgham, L., Parkes, D. C., Müller, J. and Parsons, S. (Eds), *AAMAS (2)*, pages 665–672. IFAAMAS, Richland, SC, USA.

Aldewereld, H., van der Hoek, W. and Meyer, J.-J. (2004). Rational teams: Logical aspects of multi-agent systems. *Fundamenta Informaticae*, **63**: 159–183.

Alechina, N., Bertoli, P., Ghidini, C., Jago, M., Logan, B. and Serafini, L. (2006). Model checking space and time requirements for resource-bounded agents. In *Proceedings of the Fourth International Workshop on Model Checking and Artificial Intelligence*, pages 16–30. AAAI Press, Boston (Mass), USA.

Anscombe, G. (1957). *Intention*. Cornell University Press, Ithaca, NY, USA.

Antoniou, G. (1997). *Nonmonotonic Reasoning*. MIT Press, Cambridge, MA, USA.

Arcos, J. L., Esteva, M., Noriega, P., Rodrìguez-Aguilar, J. A. and Sierra, C. (2005). Engineering open evironments with electronic institutions. *Engineering Applications of Artificial Intelligence*, **18**(2): 191–204.

Artemov, S. (2008). The logic of justification. *Review of Symbolic Logic*, **1**(4): 477–513.

Austin, J. L. (1975). *How to Do Things with Words*, Second edition, J. O. Urmson and M. Sbisa (Eds). Clarendon Press, Oxford, UK.

Balcázar, J., Dìaz, J. and Gabarró, J. (1988). *Structural Complexity I*. Springer-Verlag, New York, NY, USA.

Baltag, A., Moss, L. and Solecki, S. (2003). The logic of public announcements, common knowledge and private suspicions. Technical report, Department of Cognitive Science, Indiana University, Bloomington, IN, USA and Department of Computing, Oxford University, Oxford, UK.

Baltag, A., van Ditmarsch, H. and Moss, L. (2008). Epistemic logic and information update. In van Benthem, J. and Adriaans, P. (Eds), *Handbook on the Philosophy of Information*, Elsevier, Amsterdam, The Netherlands.

Balzer, W. and Tuomela, R. (1997). A fixed point approach to collective attitudes. In Holmstrom-Hintikka, G. and Tuomela, R. (Eds), *Contemprorary Action Theory*, Volume 2: *Social Action*, pages 115–142, Kluwer Academic Publishers, Dordrecht, The Netherlands.

Barringer, H., Kuiper, R. and Pnueli, A. (1986). A really abstract concurrent model and its temporal logic. In *POPL '86: Proceedings of the 13th ACM SIGACT-SIGPLAN Symposium on Principles of Programming Languages*, pages 173–183. ACM Press, New York, NY, USA.

Benthem, J. v. (1995). Temporal logic. In Gabbay, D., Hogger, C. J. and Robinson, J. A. (Eds), *Handbook of Logic in Artificial Intelligence and Logic Programming*, Volume 4: *Epistemic and Temporal Reasoning*, pages 241–350. Oxford University Press, Oxford, UK.

Benthem, J. v. (2001). Games in dynamic epistemic logic. *Bulletin of Economic Research*, 53(4): 219–248.

Benthem, J. v. and Pacuit, E. (2006). The tree of knowledge in action: Towards a common perspective. In Governatori, G., Hodkinson, I. M. and Venema, Y. Eds, *Advances in Modal Logic*, Volume 6, pages 87–106. College Publications London, UK.

Benthem, J. v. and Pacuit, E. (Eds) (2008). *Proceedings of the Workshop on Logic and Intelligent Interaction*. ESSLLI, Hamburg, Germany.

Benthem, J. v., van Eijck, J. and Kooi, B. (2006). Logics of communication and change. *Information and Computation*, **204**(11): 1620–1662.

Beyerlin, M. M., Johnson, D. A. and Beyerlein, S. T. (Eds) (1994). *Theories of Self-managing Work Teams*. JAI Press, Greenwich, CN, USA.

Blackburn, P., de Rijke, M. and Venema, Y. (2002). *Modal Logic*, Volume 53 of *Cambridge Tracts in Theoretical Computer Science*. Cambridge University Press, Cambridge, UK.

Blackburn, P. and Spaan, E. (1993). A modal perspective on the computational complexity of attribute value grammar. *Journal of Logic, Language and Information*, **2**: 129–169.

Boutilier, C. (1994). Toward a logic for qualitative decision theory. In Doyle, J., Sandewall, E. and Torasso, P. (Eds), *Proceedings of KR'94*, pages 75–86. Morgan Kaufmann, San Francisco, CA, USA.

Bratman, M. (1987). *Intention, Plans, and Practical Reason*. Harvard University Press, Cambridge, MA, USA.

Bratman, M. (1992). Shared cooperative activity. *The Philosophical Review*, **101**: 327–341.

Bratman, M. (1999). *Faces of Intention*. Cambridge University Press, Cambridge, UK.

Brown, M. (1988). On the logic of ability. *Journal of Philosophical Logic*, **17**: 1–26.

Brzezinski, J., Dunin-Kęcplicz, P. and Dunin-Kęcplicz, B. (2005). Collectively cognitive agents in cooperative teams. In Gleizes, M. P., Omicini, A. and Zambonelli, F. (Eds), *Engineering Societies in the Agents World V, (ESAW 2004): Revised Selected and Invited Papers*, Volume 3451 of *LNCS*, pages 191–208, Springer-Verlag, Berlin, Germany.

Cadoli, M. (1995). *Tractable Reasoning in Artificial Intelligence*. Number 941 in *Lecture Notes in Artificial Intelligence*. Springer-Verlag, Berlin, Germany.

Castelfranchi, C. (1995). Commitments: From individual intentions to groups and organizations. In Lesser, V. (Ed.), *Proceedings of the First International Conference on Multi-Agent Systems*, pages 41–48, San Francisco, CA, USA. CA, USA. AAAI Press, Menlo Park, CA, USA and MIT Press, Cambridge, MA, USA.

Castelfranchi, C. (1999). Grounding we-intentions in individual social attitudes: On social commitment again. Technical report, CNR, Institute of Psychology, Rome, Italy (manuscript).

Castelfranchi, C. (2002). The social nature of information and the role of trust. *International Journal of Cooperative Information Systems*, **11**(3): 381–403.

Castelfranchi, C. and Falcone, R. (1998). Principles of trust for MAS: Cognitive anatomy, social importance, and quantification. In Demazeau, Y. (Ed.), *Proceedings of the Third International Conference on Multi-Agent Systems*, pages 72–79, Los Alamitos, CA, USA. IEEE Computer Society, Washington, DC, USA.

Castelfranchi, C., Miceli, M. and Cesta, A. (1992). Dependence relations among autonomous agents. In Werner, E. and Demazeau, Y. (Ed.), *Decentralized A.I.-3*. Elsevier, Amsterdam, The Netherlands.

Castelfranchi, C. and Tan, Y.-H. (Ed.) (2001). *Trust and Deception in Virtual Societies*. Kluwer Academic Publishers, Dordrecht, The Netherlands.

Cavedon, L., Rao, A. and Tidhar, G. (1997). Social and individual commitment (preliminary report). In Cavedon, L., Rao, A. and Wobcke, W. (Ed.), *Intelligent Agent Systems: Theoretical and Practical Issues*, Volume 1209 of *LNAI*, pages 152–163. Springer-Verlag, Berlin, Germany.

Chapman, D. (1987). Planning for conjunctive goals. *Artificial Intelligence*, **32**: 333–377.

Chlebus, B. (1986). Domino-tiling games. *Journal of Computer and System Sciences*, **32**: 374–392.

Clark, H. H. and Marshall, C. (1981). Definite reference and mutual knowledge. In Joshi, A., Webber, B. and Sag, I. (Eds), *Elements of Discourse Understanding*, pages 10–63. Cambridge University Press, Cambridge, UK.

Cogan, E., Parsons, S. and McBurney, P. (2005). What kind of argument are we going to have today? In *Fourth International Joint Conference on Autonomous Agents and Multi Agent Systems, AAMAS 2005*, pages 544–551. ACM Press, New York, NY, USA.

Cohen, P. and Levesque, H. (1990). Intention is choice with commitment. *Artificial Intelligence*, **42**: 213–261.

Cohen, P., Levesque, H. and Smith, I. (1997). On team formation. In Holmstroem-Hintikka, G. and Tuomela, R. (Eds), *Contemporary Action Theory*, Volume II: *Social Action*, pages 87–114. Kluwer Academic Publishers, Dordrecht, The Netherlands.

Cook, S. (1971). The complexity of theorem proving procedures. In *Proceedings of the Third Annual ACM Symposium on Theory of Computing*, pages 151–158. ACM Press, New York, NY, USA.

Cuny, F. (1983). *Disasters and Development*. Oxford University Press, Oxford, UK.

d'Altan, P., Meyer, J.-J. C. and Wieringa, R. (1993). An integrated framework for ought-to-be and ought-to-do constraints. Technical Report IR-342, Department of Mathematics and Computer Science, Vrije Universiteit, Amsterdam, The Netherlands.

de Silva, L., Sardina, S. and Padgham, L. (2009). First principles planning in BDI systems. In *AAMAS '09: Proceedings of The 8th International Conference on Autonomous Agents and Multiagent Systems*, pages 1105–1112, Richland, SC, USA. International Foundation for Autonomous Agents and Multiagent Systems, Richland, SC, USA.

de Weerdt, M. M. and Clement, B. J. (2009). Introduction to planning in multiagent systems. *Multiagent and Grid Systems: An International Journal*, **5**(4): pages 345–355.

Decker, K., Zheng, X. and Schmidt, C. (2001). A multi-agent system for automated genomic annotation. In *AGENTS '01: Proceedings of the Fifth International Conference on Autonomous Agents*, pages 433–440, New York, NY, USA. ACM, New York, NY, USA.

Dennett, D. (1987). *The Intentional Stance*. MIT Press, Cambridge, MA, USA.

Denzinger, J. and Fuchs, D. (1999). Cooperation of heterogeneous provers. In Dean, T. (Ed.), *IJCAI*, pages 10–15. Morgan Kaufmann, San Francisco, CA, USA.

desJardins, M. E., Durfee, E. H., C. L. Ortiz, J. and Wolverton, M. J. (1999). A survey of research in distributed, continual planning. *AI Magazine*, **20**: 13–22.

Dignum, F. (2006). Norms and electronic institutions. In Goble, L. and Meyer, J.-J. C. (Ed.), *DEON*, Volume 4048 of *Lecture Notes in Computer Science*, pages 2–5. Springer-Verlag, Berlin, Germany.

Dignum, F. and Conte, R. (1997). Intentional agents and goal formation: Extended abstract. In Singh, M., Rao, A. and Wooldridge, M. (Eds), *Preproceedings of the Fourth International Workshop on Agent Theories, Architectures and Languages*, pages 219–231, Providence, RI, USA.

Dignum, F., Dunin-Kęplicz, B. and Verbrugge, R. (1999). Dialogue in team formation: A formal approach. In Dignum, F. and Chaib-draa, B. (Eds), *Proceedings of the IJCAI Workshop on Agent Communication Languages*, IJCAI'99. Pages 39–50, Stockholm, Sweden.

Dignum, F., Dunin-Kęplicz, B. and Verbrugge, R. (2001a). Agent theory for team formation by dialogue. In Castelfranchi, C. and Lesperance, Y. (Eds), *Intelligent Agents VII: Agent Theories, Architectures and Languages*, Volume 1986 of *Lecture Notes in Computer Science*, pages 141–156. Springer-Verlag, Berlin, Germany.

Dignum, F., Dunin-Kęplicz, B. and Verbrugge, R. (2001b). Creating collective intention through dialogue. *Logic Journal of the IGPL*, **9**: 145–158.

Dignum, F. and Kuiper, R. (1998). Specifying deadlines with continuous time using deontic and temporal logic. *International Journal of Electronic Commerce*, **3**: 67–85.

Dignum, F. and Weigand, H. (1995). Communication and deontic logic. In Wieringa, R. and Feenstra, R. (Eds), *Information Systems, Correctness and Reusability*, pages 242–260. World Scientific, Singapore.

Dignum, V. and Dignum, F. (2007). Coordinating tasks in agent organizations. In Noriega, P., Vsquez-Salceda, J., Boella, G., Boissier, O., Dignum, V., Fornara, N. and Matson, E. (Eds), *Coordination, Organizations, Institutions and Norms in Agent Systems II, Proceedings of COIN II (At AAMAS'2006 and ECAI'2006)*, Volume 4386 of *LNAI*, pages 32–47. Springer-Verlag, Berlin, Germany.

Dijksterhuis, A., Bos, M., Nordgren, L. and van Baaren, R. (2006). On making the right choice: The deliberation-without-attention effect. *Science*, **311**: 1005–1007.

d'Inverno, M., Kinny, D., Luck, M. and Wooldridge, M. (1998). A formal specification of dMARS. In *ATAL '97: Proceedings of the 4th International Workshop on Intelligent Agents IV, Agent Theories, Architectures and Languages*, pages 155–176. Springer-Verlag, London, UK.

Ditmarsch, H. v., van der Hoek, W. and Kooi, B. (2007). *Dynamic Epistemic Logic*, Volume 337 of *Synthese Library*. Springer-Verlag, Berlin, Germany.

Doherty, P., Dunin-Kęplicz, B. and Szałas, A. (2007). Dynamics of approximate information fusion. In Kryszkiewicz, M., Peters, J. F., Rybinski, H. and Skowron, A. (Eds), *RSEISP*, Volume 4585 of *Lecture Notes in Computer Science*, pages 668–677. Springer-Verlag, Berlin, Germany.

Doherty, P. and Kvarnström, J. (2008). Temporal action logics. In van Harmelen, F., Lifschitz, V. and Porter, F. (Eds), *Handbook of Knowledge Representation*. Elsevier, Amsterdam, The Netherlands.

Doherty, P., Łukaszewicz, W., Skowron, A. and Szałas, A. (2006). *Knowledge Representation Techniques. A Rough Set Approach*, Volume 202 of *Studies in Fuziness and Soft Computing*. Springer-Verlag, Berlin, Germany.

Doherty, P., Łukaszewicz, W. and Szałas, A. (2001). Computing strongest necessary and weakest sufficient conditions of first-order formulas. *International Joint Conference on AI (IJCAI'2001)*, pages 145–151. Morgan Kaufmann Publishers, San Francisco, CA, USA.

Doherty, P., Łukaszewicz, W. and Szałas, A. (2003). On mutual understanding among communicating agents. In Dunin-Kęplicz, B. and Verbrugge, R. (Eds), *Proceedings of Workshop on Formal Approaches to Multi-agent Systems (FAMAS'03)*, pages 83–97.

Downey, R. G. and Fellows, M. R. (1995). Fixed-parameter tractability and completeness I: Basic results. *SIAM Journal of Computing*, **24**(4): 873–921.

Drimmelen, G. v. (2003). Satisfiability in alternating-time temporal logic. In *LICS '03: Proceedings of the 18th Annual IEEE Symposium on Logic in Computer Science*, page 208. IEEE Computer Society, Washington, DC, USA.

Dunin-Kęplicz, B., Nguyen, A. and Szałas, A. (2010). Tractable approximate knowledge fusion using the Horn fragment of serial propositional dynamic logic. *International Journal of Approximate Reasoning*, **51**(3): 346–362.

Dunin-Kęplicz, B., Nguyen, L. and Szałas, A. (2009a). Fusing approximate knowledge from distributed sources. In Papadopoulos, G. and Badica, C. (Eds), *Proceedings of the Third International Symposium on Intelligent Distributed Computing*, Volume 237 of *Studies in Computational Intelligence*, pages 75–86. Springer-Verlag, Berlin.

Dunin-Kęplicz, B. and Szałas, A. (2007). Towards approximate BGI systems. In Burkhard H. D., Lindemann, G., Verbrugge, R. and Varga, Z. L. (Eds), *CEEMAS*, Volume 4696 of *Lecture Notes in Computer Science*, pages 277–287. Springer-Verlag, Berlin, Germany.

Dunin-Kęplicz, B. and Szałas, A. (2010). Agents in approximate environments. In van Eijck, J. and Verbrugge, R. (Eds), *Games, Actions and Social Software*. College Publications, London, UK.

Dunin-Kęplicz, B., Verbrugge, R. and Slizak, M. (2009b). Case-study for TeamLog, a theory of teamwork. In Papadopoulos, G. and Badica, C. (Eds), *Proceedings of the Third International Symposium on Intelligent Distributed Computing*, Volume 237 of *Studies in Computational Intelligence*, pages 87–100. Springer-Verlag, Berlin.

Dunin-Kęplicz, B. and Radzikowska, A. (1995a). Actions with typical effects: Epistemic characterization of scenarios. In Lesser, V. (Ed.), *Proceedings of the First International Conference on Multi-Agent Systems*, page 445, San Francisco, San Francisco, CA, USA. AAAI-Press, Menlo Park, CA, USA and MIT Press, Cambridge, MA, USA.

Dunin-Kęplicz, B. and Radzikowska, A. (1995b). Epistemic approach to actions with typical effects. In *Proceedings ECSQARU'95*, pages 180–189, Fribourg, Switzerland.

Dunin-Kęplicz, B. and Radzikowska, A. (1995c). Modelling nondeterminstic actions with typical effects. In *Proceedings DIMAS'95*, pages 158–166, Cracow, Poland.

Dunin-Kęplicz, B. and Verbrugge, R. (1996). Collective commitments. In Tokoro, M. (Ed.), *Proceedings of the Second International Conference on Multi-Agent Systems*, pages 56–63. AAAI-Press, Menlo Park, CA, USA.

Dunin-Kęplicz, B. and Verbrugge, R. (1999). Collective motivational attitudes in cooperative problem solving. In Gorodetsky, V. (Ed.), *Proceedings of the First International Workshop of Eastern and Central Europe on Multi-agent Systems (CEEMAS'99)*, pages 22–41, St. Petersburg, Russia.

Dunin-Kęplicz, B. and Verbrugge, R. (2001a). A reconfiguration algorithm for distributed problem solving. *Engineering Simulation*, **18**: 227–246.

Dunin-Kęplicz, B. and Verbrugge, R. (2001b). The role of dialogue in collective problem solving. In Davis, E., McCarthy, J., Morgenstern, L. and Reiter, R. (Eds), *Proceedings of the Fifth International Symposium on the Logical Formalization of Commonsense Reasoning (Commonsense 2001)*, pages 89–104, New York, NY, USA.

Dunin-Kęplicz, B. and Verbrugge, R. (2002). Collective intentions. *Fundamenta Informaticae*, **51**(3): 271–295.

Dunin-Kęplicz, B. and Verbrugge, R. (2004). A tuning machine for cooperative problem solving. *Fundamenta Informaticae*, **63**: 283–307.

Dunin-Kęplicz, B. and Verbrugge, R. (2005). Creating common beliefs in rescue situations. In Dunin-Kęplicz, B., Jankowski, A., Skowron, A. and Szczuka, M. (Eds), *Proceedings of Monitoring, Security and Rescue Techniques in Multiagent Systems (MSRAS)*, Advances in Soft Computing, pages 69–84. Springer-Verlag, Berlin, Germany.

Dunin-Kęplicz, B. and Verbrugge, R. (2006). Awareness as a vital ingredient of teamwork. In Stone, P. and Weiss, G. (Eds), *Proceedings of the Fifth International Joint Conference on Autonomous Agents and Multi-agent Systems (AAMAS'06)*, pages 1017–1024, New York (NY). IEEE Computer Press, Washington, DC, USA and ACM Press, New York, NY, USA.

Durfee, E. H. (2008). Planning for coordination and coordination for planning. In *Web Intelligence and Intelligent Agent Technology, IEEE/WIC/ACM International Conference on*, Volume 1, pages 1–3, Los Alamitos, CA, USA. IEEE Computer Press, Washington, DC, USA.

Dziubiński, M. (2007). Complexity of the logic for multiagent systems with restricted modal context. In Dunin-Kęplicz, B. and Verbrugge, R. (Eds), *Proceedings of the Third Workshop on Formal Approaches to Multi-agent Systems (FAMAS'007)*, pages 1–18, Durham. Durham University, Press, Durham, UK.

Dziubiński, M. (in preparation). *Complexity Issues in Multimodal Logics for Multiagent Systems*. PhD thesis, Institute of Computer Science, University of Warsaw, Poland.

Dziubiński, M., Verbrugge, R. and Dunin-Kęplicz, B. (2007). Complexity issues in multiagent logics. *Fundamenta Informaticae*, **75**(1–4): 239–262.

El Fallah-Seghrouchni, A. (1997). Multi-agent systems as a paradigm for intelligent system design. *Informatica (Slovenia)*, **21**(2).

Emerson, A. (1990). Temporal and modal logic. In van Leeuwen, J. (Ed.), *Handbook of Theoretical Computer Science*, Volume B, pages 995–1072. Elsevier, Amsterdam, The Netherlands and MIT Press, Cambridge, MA, USA.

Engelmore, R. and Morgan, A. (1988). *Blackboard Systems*. Addison-Wesley, New York, NY, USA.

Fagin, R. and Halpern, J. (1988). Belief, awareness, and limited reasoning. *Artificial Intelligence*, **34**(1): 39–76.

Fagin, R., Halpern, J., Moses, Y. and Vardi, M. (1995). *Reasoning about Knowledge*. MIT Press, Cambridge, MA, USA.

Ferber, J. (1999). *Multi-agent Systems: An Introduction to Distributed Artificial Intelligence*. Addison-Wesley, Reading, MA, USA.

Ferguson, I. A. (1992). Touring machines: Autonomous agents with attitudes. *Computer*, **25**(5): 51–55.

Fischer, M. and Ladner, R. (1979). Propositional dynamic logic of regular programs. *Journal of Computer and System Sciences*, **18**(2): 194–211.

Fisher, M. (1994). A survey of concurrent METATEM – the language and its applications. In Gabbay, D. M. and Ohlbach, H. J. (Eds), *Temporal Logic - Proceedings of the First International Conference*, Volume 827 of *LNAI*, pages 480–505. Springer-Verlag, Heidelberg, Germany.

Flobbe, L., Verbrugge, R., Hendriks, P. and Krämer, I. (2008). Children's application of theory of mind in reasoning and language. *Journal of Logic, Language and Information*, **17**: 417–442. Special issue on formal models for real people, edited by M. Counihan.

Fortnow, L. (2009). The status of the P versus NP problem. *Communications of the ACM*, **52**(9): 78–86.

Foss, J. N. and Onder, N. (2006). A hill-climbing approach for planning with temporal uncertainty. In Sutcliffe, G. and Goebel, R., (Eds), *FLAIRS Conference*, pages 868. AAAI Press, Menlo Park, CA, USA.

Gabbay, D. (1977). Axiomatization of logic programs. Technical report. Text of letter to V. Pratt.

Gabbay, D. (1998). *Fibring Logics*, Volume 38 of *Oxford Logic Guides*. Oxford Univesity Press, Oxford, UK.

Gabbay, D. M., Schmidt, R. A. and Szałas, A. (2008). *Second-Order Quantifier Elimination: Foundations, Computational Aspects and Applications*, Volume 12 of *Studies in Logic: Mathematical Logic and Foundations*. College Publications, London, UK.

Gasser, L. (1995). Computational organization research. In Lesser, V. R. and Gasser, L. (Eds), *ICMAS*, pages 414–415. MIT Press, Cambridge, MA, USA.

Georgeff, M. and Lansky, A. (1987). Reactive reasoning and planning. In *AAAI-87 Proceedings*, pages 677–682. AAAI Press, Menlo Park, CA, USA.

Gettier, E. (1963). Is justified true belief knowledge? *Analysis*, **23**: 121–123.

Gilbert, M. (1989). *On Social Facts*. International Library of Philosophy. Routledge (Taylor & Francis), Abingdon, UK.

Gilbert, M. (2005). A theoretical framework for the understanding of teams. In Gold, N. (Ed.), *Teamwork*, pages 22–32. Palgrave McMillan, Basingstoke, UK and New York, NY, USA.

Gilbert, M. (2009). Shared intention and personal intentions. *Philosophical Studies*, **144**: 1167–187.

Gochet, P. and Gillet, E. (1991). On Professor Weingartner's contribution to epistemic logic. In *Advances in Scientific Philosophy*, pages 97–115. Rodopi, Amsterdam, The Netherlands.

Gold, N. (Ed.) (2005). *Teamwork*. Palgrave McMillan, Basingstoke, UK and New York, NY, USA.

Goldblatt, R. (1992). *Logics of Time and Computation*. Number 7 in CSLI Lecture Notes. Center for Studies in Language and Information, Palo Alto, CA, USA.

Gonzalez, A. and Dankel, D. (1993). *The Engineering of Knowledge-based Systems: Theory and Practice*. Prentice-Hall, Upper Saddle River, NJ, USA.

Goranko, V. and Jamroga, W. (2004). Comparing semantics of logics for multi-agent systems. *Synthese*, **139**(2): 241–280.

Goranko, V. and Shkatov, D. (2009). Tableau-based decision procedure for full coalitional multiagent temporal-epistemic logic of branching time. In Baldoni, M., Dunin-Kęplicz, B. and Verbrugge, R. (Eds), *Proceedings of the 2nd Federated Workshops on Multi-Agent Logics, Languages and Organisations, MALLOW'009*, Turin, Italy. CEUR Workshop Proceedings, Vol. 494.

Graedel, E. (1999). Why is modal logic so robustly decidable? *Bulletin of the EATCS*, **68**: 90–103.

Grant, J., Kraus, S. and Perlis, D. (2005a). Formal approaches to teamwork. In Artemov, S., Barringer, H., d'Avila Garcez, A. S., Lamb, L. C. and Woods, J. (Eds), *We Will Show Them: Essays in Honour of Dov Gabbay*, Volume 1, pages 39–68. College Publications, London.

Grant, J., Kraus, S. and Perlis, D. (2005b). A logic-based model of intention formation and action for multi-agent subcontracting. *Artificial Intelligence*, **163**(2): 163–201.

Grossi, D., Royakkers, L. and Dignum, F. (2007). Organizational structure and responsibility. an analysis in a dynamic logic of organized collective agency. *Journal of AI and Law*, **15**(3): 223–249.

Grosz, B. and Kraus, S. (1996). Collaborative plans for complex group action. *Artificial Intelligence*, **86**(2): 269–357.

Grosz, B. and Kraus, S. (1999). The evolution of SharedPlans. In Rao, A. and Wooldridge, M. (Eds), *Foundations of Rational Agency*, pages 227–262. Kluwer Academic Publishers, Dordrecht, The Netherlands.

Grosz, B. and Sidner, C. (1990). Plans for discourse. In Cohen, P., Morgan, J. and Pollack, M. (Eds), *Intentions in Communication*, pages 417–444. MIT Press, Cambridge, MA, USA.

Haddadi, A. (1995). *Communication and Cooperation in Agent Systems: A Pragmatic Theory*, Volume 1056 of *LNAI*. Springer-Verlag, Berlin, Germany.

Hallett, M. T. and Wareham, H. T. (1994). The parameterized complexity of some problems in logic and linguistics. In Nerode, A. and Matiyasevich, Y. (Eds), *Logical Foundations of Computer Science, Third International Symposium, LFCS'94*, Volume 813 of *Lecture Notes in Computer Science*, pages 89–100. Springer-Verlag, Berlin, Germany.

Halpern, J. (1996). Should knowledge entail belief? *Journal of Philosophical Logic*, **25**: 483–494.

Halpern, J. and Zuck, L. (1987). A little knowledge goes a long way: Simple knowledge-based derivations and correctness proofs for a family of protocols. In *6th ACM Symposium on Principles of Distributed Computing*, pages 268–280.

Halpern, J. Y. (1995). The effect of bounding the number of primitive propositions and the depth of nesting on the complexity of modal logic. *Artificial Intelligence*, **75**(3): 361–372.

Halpern, J. Y. and Moses, Y. (1984). Knowledge and common knowledge in a distributed environment. In *Symposium on Principles of Distributed Computing*, pages 50–61.

Halpern, J. Y. and Moses, Y. (1992). A guide to completeness and complexity for modal logics of knowledge and belief. *Artificial Intelligence*, **54**(3): 319–379.

Harbers, M., Verbrugge, R., Sierra, C. and Debenham, J. (2008). The examination of an information-based approach to trust. In Sichman, J., Padget, J., Ossowski, S. and Noriega, P. (Eds), *Coordination, Organizations, Institutions and Norms in Agent Systems III*, Volume 4870 of *Lecture Notes in Computer Science*, pages 71–82. Springer-Verlag, Berlin, Germany.

Harel, D., Kozen, D. and Tiuryn, J. (2000). *Dynamic Logic*. MIT Press, Cambridge, MA, USA.

Hedden, T. and Zhang, J. (2002). What do you think I think you think? Strategic reasoning in matrix games. *Cognition*, **85**: 1–36.

Hintikka, J. (1962). *Knowledge and Belief*. Cornell University Press, Ithaca, NY, USA.

Hintikka, J. (1973). *Logic, Language-Games and Information: Kantian Themes in the Philosophy of Logic*. Clarendon Press, Oxford, UK.

Hustadt, U. and Schmidt, R. (1997). On evaluating decision procedures for modal logics. In Pollack, M. (Ed.), *Proceedings IJCAI'97*, Los Angeles, CA, USA. Morgan Kauffman, San Francisco, CA, USA.

Jamroga, W. and van der Hoek, W. (2004). Agents that know how to play. *Fundamenta Informaticae*, **63**(2–3): 185–219.

Jennings, N. (1993). Commitments and conventions: The foundation of coordination in multi-agent systems. *Knowledge Engineering Review*, **3**: 223–250.

Jennings, N. R. and Bussmann, S. (2003). Agent-based control systems: Why are they suited to engineering complex systems? *IEEE Control Systems Magazine*, **23**(3): 61–74.

Jennings, N. R., Sycara, K. and Wooldridge, M. (1998). A roadmap of agent research and development. *Autonomous Agents and Multi-agent Systems*, **1**: 7–38.

Jennings, N. R. and Wooldridge, M. (2000). Agent-oriented software engineering. *Artificial Intelligence*, **117**: 277–296.

Jøsang, A., Ismail, R. and Boyd, C. (2007). A survey of trust and reputation systems for online service provision. *Decision Support Systems*, **43**(2): 618–644.

Kacprzak, M., Lomuscio, A. and Penczek, W. (2004). From bounded to unbounded model checking for interpreted systems. *Fundamenta Informaticae*, **63**(2–3): 107–308.

Kambartel, F. (1979). Constructive pragmatics and semantics. In Bauerle, R. (Ed.), *Semantics from a Different Point of View*, pages 195–205. Springer-Verlag, Berlin, Germany.

Kautz, H. and Selman, B. (1996). Knowledge compilation and theory approximation. *Journal of the ACM*, **43**(2): 193–224.

Keysar, B., Lin, S. and Barr, D. (2003). Limits on theory of mind use in adults. *Cognition*, **89**: 25–41.

Kleiner, A., Prediger, J. and Nebel, B. (2006). RFID technology-based exploration and slam for search and rescue. In *Proceedings of the IEEE/RSJ International Conference on Intelligent Robots and Systems (IROS 2006)*, pages 4054–4059, Bejing, China.

Konolige, K. and Pollack, M. E. (1993). A representationalist theory of intention. In Bajcsy, R. (Ed.), *Proceedings of the Thirteenth International Joint Conference on Artificial Intelligence (IJCAI-93)*, pages 390–395, Chambéry, France. Morgan Kaufmann, San Mateo, CA, USA.

Koza, J., Keane, M., Mydlowec, W., Yu, J. and Lanza, G. (2003). *Genetic Programming IV: Routine Human-Competitive Machine Intelligence*. Kluwer Academic Publishers, Dordrecht, The Netherlands.

Kozen, D. and Parikh, R. (1981). An elementary proof of the completeness of PDL. *Theoretical Computer Science*, **14**: 113–118.

Krabbe, E. (2001). Dialogue foundations: Dialogue logic revisited. *Supplement to the Proceedings of The Aristotelian Society*, **75**: 33–49.

Kraus, S. (2001). *Strategic Negotiation in Multiagent Environments*. MIT Press, Cambridge, MA, USA.

Kraus, S. and Lehmann, D. (1988). Knowledge, belief and time. *Theoretical Computer Science*, **58**: 155–174.

Lao-Tzu (1992). *Tao Te Ching*. Harper Perennial, New York, NY, USA. English version translated by Stephen Mitchell.

Lathrop, R. (1994). The protein threading problem with sequence amino acid interaction preferences is NP-complete. *Protein Engineering*, **7**: 1059–1068.

Lemmon, E. (1977). *An Introduction to Modal Logic (the Lemmon Notes)*. Number 1 in Monograph Series, *American Philosophical Quarterly*, written in collaboration with D. Scott, edited by K. Segerberg. University of Illinois Press, Champaign, IL, USA.

Lenzen, W. (1978). Recent works in epistemic logic. *Acta Philosophica Fennica*, **30**: 1–219.

Levesque, H., Cohen, P. and Nunes, J. (1990). On acting together. In *Proceedings of the Eighth National Conference on AI (AAAI90)*, pages 94–99. AAI-Press, Menlo Park, CA, USA and MIT Press, Cambridge, MA, USA.

Lewis, C. and Langford, C. (1959). *Symbolic Logic*, 2nd Edition. Dover, New York, NY, USA.

Lewis, D. (1969). *Convention: A Philosophical Study*. Harvard University Press, Cambridge, MA, USA.

Lin, F. (2000). On strongest necessary and weakest sufficient conditions. In Cohn, A., Giunchiglia, F. and Selman, B. (Eds), *Proceedings of the 7th International Conference on Principles of Knowledge Representation and Reasoning, KR2000*, pages 167–175. Morgan Kaufmann, San Francisco, CA, USA.

Lin, R. and Kraus, S. (2008). Negotiating with bounded rational agents in environments with incomplete information using an automated agent. *Artificial Intelligence Journal*, **172**(6–7): 823–851.

Litman, L. and Reber, A. (2005). Implicit cognition and thought. In Holyoak, K. and Morrison, R. (Eds), *The Cambridge Handbook of Thinking and Reasoning*, pages 431–453. Cambridge University Press, Cambridge, UK.

Lorenzen, P. (1961). Ein dialogisches Konstruktivitaetskriterium. In *Infinitistic Methods*, pages 193–200. Pergamon Press, Oxford, UK.

Luck, M., McBurney, P. and Preist, C. (2003). *Agent Technology: Enabling Next Generation Computing: A Roadmap for Agent Based Computing*. Agentlink.

Łukaszewicz, W. (1990). *Non-Monotonic Reasoning*. Ellis-Horwood, Chichester, UK.

Marsh, S. and Dibben, M. R. (2003). The role of trust in information science and technology. *Annual Review of Information Science and Technology*, **37**(1): 465–498.

McBurney, P., Hitchcock, D. and Parsons, S. (2007). The eightfold way of deliberation dialogue. *International Journal of Intelligent Systems*, **22**(1): 95–132.

McBurney, P., Parsons, S. and Wooldridge, M. (2002). Desiderata for agent argumentation protocols. In *First International Joint Conference on Autonomous Agents and Multi Agent Systems, AAMAS 2002*, pages 402–409. ACM Press, New York, NY, USA.

McCarthy, J. (1986). Applications of circumscription to formalizing common-sense knowledge. *Artificial Intelligence*, **28**(1): 89–116.

McCarthy, J. and Hayes, P. J. (1969). Some philosophical problems from the standpoint of artificial intelligence. *Machine Intelligence*, **4**: 463–502.

Meyer, J.-J. C. and van der Hoek, W. (1995). *Epistemic Logic for AI and Theoretical Computer Science*. Cambridge University Press, Cambridge, UK.

Mirkowska, G. and Salwicki, A. (1987). *Algorithmic Logic*. Kluwer Academic Publishers, Norwell, MA, USA.

Mol, L., Taatgen, N., Verbrugge, L. and Hendriks, P. (2005). Reflective cognition as secondary task. In Bara, B., Barsalou, L. and Bucciarelli, M. (Eds), *Proceedings of the Twenty-seventh Annual Meeting of the Cognitive Science Society*, pages 1925–1930. Lawrence Erlbaum Publishers, Mahwah, NJ, USA.

Montague, R. (1973). The proper treatment of quantification in ordinary english. In Hintikka, J., Moravcsik, J. and Suppes, P. (Eds), *Approaches to Natural Language: Proceedings of the 1970 Stanford Workshop on Grammar and Semantics*, pages 221–242. Reidel, Dordrecht, The Netherlands.

Müller, J. P. (1997). A cooperation model for autonomous agents. In *ECAI '96: Proceedings of the Workshop on Intelligent Agents III, Agent Theories, Architectures and Languages*, pages 245–260, London, UK. Springer-Verlag, Berlin, Germany.

Nair, R., Tambe, M. and Marsella, S. (2003). Team formation for reformation in multiagent domains like roboCupRescue. In Kaminka, G. A., Lima, P. U. and Rojas, R. (Eds), *RoboCup*, Volume 2752 of *Lecture Notes in Computer Science*, pages 150–161. Springer-Verlag, Berlin, Germany.

Nebel, B. and Koehler, J. (1992). A validation-structure-based theory of plan modification and reuse. *Artificial Intelligence*, **55**: 193–258.

Nebel, B. and Koehler, J. (1995). Plan reuse versus plan generation: A theoretical and empirical analysis. *Artificial Intelligence*, **76**(1–2): 427–454.

Nguyen, L. A. (2005). On the complexity of fragments of modal logics. In *Advances in Modal Logic – Volume 5*, pages 249–268. King's College Publications, London.

Niewiadomski, A., Penczek, W. and Szreter, M. (2009). A new approach to model checking of UML state machines. *Fundamenta Informormaticae*, **93**(1–3): 289–303.

Papadimitriou, C. (1993). *Computational Complexity*. Addison-Wesley, Reading, MA, USA.

Parikh, R. (1978). The completeness of propositional dynamic logic. In Winkowski, J. (Ed.), *MFCS*, Volume 64 of *Lecture Notes in Computer Science*, pages 403–415. Springer-Verlag, Berlin, Germany.

Parikh, R. (2002). States of knowledge. *Electronic Notes in Theoretical Computer Science*, Volume 67.

Parikh, R. (2005). Logical omniscience and common knowledge: WHAT do we know and what do WE know? In *TARK '05: Proceedings of the 10th Conference on Theoretical aspects of Rationality and Knowledge*, pages 62–77, Singapore. National University of Singapore, Singapore.

Parikh, R. and Krasucki, P. (1992). Levels of knowledge in distributed computing. *Sadhana: Proceedings of the Indian Academy of Sciences*, **17**: 167–191.

Parsons, S., Wooldridge, M. and Amgoud, L. (2003). On the outcomes of formal inter-agent dialogues. In *Second International Joint Conference on Autonomous Agents and Multi Agent Systems, AAMAS 2003*, pages 616–623. ACM Press, New York, NY, USA.

Parsons, S., Wooldridge, M. and Jennings, N. (1998). Agents that reason and negotiate by arguing. *Journal of Logic and Computation*, **8**: 261–292.

Pawlak, Z. (1981). Information systems – theoretical foundations. *Information Systems*, **6**: 205–218.

Pawlak, Z. (1991). *Rough Sets. Theoretical Aspects of Reasoning about Data*. Kluwer Academic Publishers, Dordrecht, The Netherlands.

Peleg, D. (1987). Concurrent dynamic logic. *Journal of the ACM*, **34**(2): 450–479.

Penczek, W. and Lomuscio, A. (2003). Verifying epistemic properties of multi-agent systems via bounded model checking. *Fundamenta Informaticae*, **55**(2). 167–185.

Penczek, W. and Szreter, M. (2008). SAT-based unbounded model checking of timed automata. *Fundamenta Informaticae*, **85**(1–4): 425–440.

Procaccia, A. D. and Rosenschein, J. S. (2006). The communication complexity of coalition formation among autonomous agents. In Nakashima, H., Wellman, M.P., Weiss, G. and Stone, P. (Eds), *The Fifth International Joint Conference on Autonomous Agents and Multiagent Systems*, pages 505–512, Hakodate, Japan.

Purser, R. and Cabana, S. (1998). *The Self-managing Organization*. Free Press, New York, NY, USA.

Pynadath, D. V. and Tambe, M. (2002). The communicative multiagent team decision problem: Analyzing teamwork theories and models. *Journal of Artificial Intelligence Research*, **16**: 389–423.

Quine, W. (1956). Quantifiers and propositional attitudes. *Journal of Philosophy*, **53**: 177–187.

Ramchurn, S. D., Huynh, D. and Jennings, N. R. (2004). Trust in multiagent systems. *Knowledge Engineering Review*, **19**(1): 1–25.

Ramchurn, S. D., Sierra, C., Godo, L. and Jennings, N. R. (2007). Negotiating using rewards. *Artificial Intelligence*, **171**: 805–837.

Rao, A. (1996). Agentspeak(l): Agents speak out in a logical computable language. In Van de Velde, W. and Perram, J. (Eds), *Agents Breaking Away: Proceedings of the 7th European Workshop on Modelling Autonomous Agents in a Multi-Agent World, MAAMAW'96*, Volume 1038 of *LNCS*, pages 42–55. Springer-Verlag, Berlin, Germany.

Rao, A. and Georgeff, M. (1991). Modeling rational agents within a BDI-architecture. In Fikes, R. and Sandewall, E. (Eds), *Proceedings of the Second Conference on Knowledge Representation and Reasoning*, pages 473–484. Morgan Kaufman, San Francisco, CA, USA.

Rao, A. and Georgeff, M. (1995a). BDI agents: From theory to practice. In Lesser, V. (Ed.), *Proceedings of the First International Conference on Multi-Agent Systems*, pages 312–319, San Francisco, CA, USA. AAAI-Press, Menlo Park, CA, USA and MIT Press, Cambridge, MA, USA.

Rao, A. and Georgeff, M. (1995b). Formal models and decision procedures for multi-agent systems. Technical Report Technical Note 61. Australian Artificial Intelligence Institute, Carlton, Victoria, Australia.

Rao, A., Georgeff, M. and Sonenberg, E. (1992). Social plans: A preliminary report. In Werner, E. and Demazeau, Y. (Eds), *Decentralized A.I.-3*, pages 57–76. Elsevier, Amsterdam, The Netherlands.

Reed, C. (1998). Dialogue frames in agent communication. In Demazeau, Y. (Ed.), *Proceedings of the Third International Conference on Multi-Agent Systems*, pages 246–253, Los Alamitos, CA, USA. IEEE Computer Society, Washington, DC, USA.

Reiter, R. (1980). A logic for default reasoning. *Artificial Intelligence*, **13**: 81–132.

Reiter, R. (1991). The frame problem in the situation calculus: a simple solution (sometimes) and a completeness result for goal regression. In Lifschitz, V. (Ed.), *Artificial Intelligence and Mathematical Theory of Computation: Papers in Honor of John McCarthy*, pages 359–380. Academic Press, New York, NY, USA.

Rosenschein, J. and Zlotkin, G. (1994). *Rules of Encounter: Designing Conventions for Automated Negotiation Among Computers*. MIT Press, Cambridge, MA, USA.

Roy, O. (2006). Commitment-based decision making for bounded agents. In Ågotnes, T. and Alechina, N. (Eds), *Proceedings of the Workshop on Resource-bounded Agents*, pages 112–123, European Summer School on Logic, Language and Information, Malaga, Spain.

Salwicki, A. (1970). Formalized algorithmic languages. *Bulletin de l'Académie Polonaise des Sciences: Série des Sciences Mathématiques, Astronomiques et Physiques*, **18**: 227–232.

Sandewall, E. (1994). *Features and Fluents*. Oxford University Press, Oxford, UK.

Sandholm, T. and Lesser, V. (1995). Issues in automated negotiation and electronic commerce: extending the Contract Net Protocol. In Lesser, V. (Ed.), *Proceedings of the First International Conference on Multi-Agent Systems*, pages 328–335, San Francisco, CA, USA. AAAI-Press, Menlo Park, CA, USA and MIT Press, Cambridge, MA, USA.

Savitch, W. (1970). Relationship between nondeterministic and deterministic tape classes. *Journal of Computer and System Sciences*, **4**: 177–192.

Searle, J. and Vanderveken, D. (1985). *Foundations of Illocutionary Logic*. Cambridge University Press, Cambridge, UK.

Searle, J. R. (1969). *Speech Acts*. Cambridge University Press, Cambridge, UK.

Segerberg, K. (1977). A completeness theorem in the modal logic of programs. *Notices of the AMS*, **24**(6): A–552.

Segerberg, K. (1989). Bringing it about. *Journal of Philosophical Logic*, **18**: 327–347.

Shehory, O. (2004). Coalition formation: Towards feasible solutions. *Fundamenta Informaticae*, **63**(2–3): 107–124.

Shehory, O. and Kraus, S. (1998). Methods for task allocation via agent coalition formation. *Artificial Intelligence*, **101**(1–2): 165–200.

Sichman, J. S. and Conte, R. (2002). Multi-agent dependence by dependence graphs. In *AAMAS*, pages 483–490. ACM Press, New York, NY, USA.

Sierra, C. and Debenham, J. (2007). The LOGIC negotiation model. In *Sixth International Joint Conference on Autonomous Agents and Multi-agent Systems (AAMAS'2007)*, pages 1026–1033. IFAAMAS, Honolulu, Hawaii, USA.

Singh, M. (1990). Group intentions. In *Proceedings of the 10th International Workshop on Distributed Artificial Intelligence*.

Singh, M. (1997). Commitments among autonomous agents in information-rich environments. In Boman, M. and de Velde, W. V. (Eds), *Multi-Agent Rationality (Proceedings of MAAMAW'97)*, Volume 1237 of *LNAI*, pages 141–155. Springer-Verlag, Berlin, Germany.

Singh, M. (1998). The intentions of teams. In Prade, H. (Ed.), *Proceedings of the 13th European Conference on Artificial Intelligence (ECAI'98)*, pages 303–307. John Wiley & Sons, Ltd, Chichester, UK.

Sonenberg, E., Tidhar, G., Werner, E., Kinny, D., Ljungberg, M. and Rao, A. (1992). Planned team activity. In *Artificial Social Systems*, Volume 890 of *LNAI*. Springer-Verlag, Berlin, Germany.

Stefik, M. (1995). *Introduction to Knowledge Systems*. Morgan Kaufmann, San Francisco, CA, USA.

Stulp, F. and Verbrugge, R. (2002). A knowledge-based algorithm for the internet protocol TCP. *Bulletin of Economic Research*, **54**(1): 69–94.

Sycara, K. (1990). Persuasive argumentation in negotiation. *Theory and Decision*, **28**: 203–242.

Sycara, K. and Lewis, M. (2004). Integrating intelligent agents into human teams. In Salas, E. and Fiore, S. (Eds), *Team Cognition: Understanding the Factors that Drive Process and Performance*, pages 203–232. American Psychological Association, Washington, DC, USA.

Szałas, A. (1995). Temporal logic: A standard approach. In Bolc, L. and Szałas, A. (Eds), *Time and Logic. A Computational Approach*, pages 1–50. UCL Press, London, UK.

Tambe, M. (1996). Teamwork in real-world, dynamic environments. In Tokoro, M. (Ed.), *Proceedings of the Second International Conference on Multi-Agent Systems*, pages 361–368, Menlo Park, CA, USA. AAA-I Press, Menlo Park, CA, USA.

Tambe, M. (1997). Towards flexible teamwork. *Journal of Artificial Intelligence Research*, **7**: 83–124.

Traum, D. (1999). Speech acts for dialogue agents. In Rao, A. and Wooldridge, M. (Eds), *Foundations of Rational Agency*, pages 169–201. Kluwer Academic Publishers, Dordrecht, The Netherlands.

Tuomela, R. (1995). *The Importance of Us: A Philosophical Study of Basic Social Notions*. Stanford Series in Philosophy. Stanford University Press, Stanford, CA, USA.

Tuomela, R. and Miller, K. (1988). We-intentions. *Philosophical Studies*, **53**: 367–390.

Van Baars, E. and Verbrugge, R. (2007). Adjusting a knowledge-based algorithm for multi-agent communication for cps. In Dastani, M., Fallah-Seghrouchni, A. E., Leite, J. and Torroni, P. (Eds), *LADS*, Volume 5118 of *Lecture Notes in Computer Science*, pages 89–105. Springer-Verlag, Berlin, Germany.

van Benthem, J. (2005). Correspondence theory. In Gabbay, D. and Guenthner, F. (Eds), *Handbook of Philosophical Logic* (Second Edition), Volume 3, pages 325–408. Kluwer Academic Publishers, Dordrecht, The Netherlands. An earlier version appeared in volume II of the First Edition of the Handbook.

van Benthem, J. and Liu, F. (2007). Dynamic logic of preference upgrade. *Journal of Applied Non-Classical Logics*, **17**(2): 157–182.

van der Hoek, W., Jamroga, W. and Wooldridge, M. (2007). Towards a theory of intention revision. *Synthese*, **155**(2): 265–290.

van der Hoek, W. and Verbrugge, R. (2002). Epistemic logic: A survey. In Petrosjan, L. and Mazalov, V. (Eds), *Game Theory and Applications*, Volume 8 pages 53–94. Nova Science Publishers, New York, NY, USA.

van Ditmarsch, H. and Kooi, B. (2003). Unsuccessful updates. In Álvarez, E., Bosch, R., and Villamil, L. (Eds), *Proceedings of the 12th International Congress of Logic, Methodology, and Philosophy of Science (LMPS)*, pages 139–140. Oviedo University Press, Oviedo, Spain.

van Linder, B., van der Hoek, W. and Meyer, J.-J. C. (1998). Formalising abilities and opportunities of agents. *Fundamenta Informaticae*, **34**: 53–101.

van Rooij, I. (2008). The tractable cognition thesis. *Cognitive Science*, **32**: 939–984.

Vardi, M. (1997). Why is modal logic so robustly decidable? *DIMACS Series on Discrete Mathematics and Theoretical Computer Science*, **31**: 149–184.

Velleman, D. (2000). *The Possibility of Practical Reason*. Oxford University Press, Oxford, UK.

Verbrugge, R. and Mol, L. (2008). Learning to apply theory of mind. *Journal of Logic, Language and Information*, **17**: 489–511. Special issue on formal models for real people, edited by M. Counihan.

Voorbraak, F. (1991). The logic of objective knowledge and rational belief. In *JELIA '90: Proceedings of the European workshop on Logics in AI*, pages 499–515, New York, NY, USA. Springer-Verlag, New York, NY, USA.

Walther, D., Lutz, C., Wolter, F. and Wooldridge, M. (2006). ATL satisfiability is indeed EXPTIME-complete. *Journal of Logic and Computation*, **16**(6): 765–787.

Walton, D. and Krabbe, E. (1995). *Commitment in Dialogue: Basic Concepts of Interpersonal Reasoning*. State University of New York Press, Albany, NY, USA.

Weiss, G. (Ed.) (1999). *Multiagent Systems*. MIT Press, Cambridge, MA, USA.

Williamson, O. (1981). The economies of organization: The transaction cost approach. *American Journal of Sociology*, **87**: 548–577.

Wisner, B., Blaikie, P., Cannon, T. and Davis, I. (2004). *At Risk – Natural Hazards, People's Vulnerability and Disasters*. Routledge (Taylor & Francis), Abingdon, UK.

WITAS (2001). *Project web page* [http://www.ida.liu.se/ext/witas/eng.html].

Wooldridge, M. (2000). *Reasoning About Rational Agents*. MIT Press, Cambridge, MA, USA.

Wooldridge, M. (2009). *MultiAgent Systems*, Second Edition. John Wiley & Sons, Ltd, Chichester, UK.

Wooldridge, M. and Jennings, N. (1996). Towards a theory of collective problem solving. In Perram, J. and Muller, J. (Eds), *Distributed Software Agents and Applications*, Volume 1069 of *LNAI*, pages 40–53. Springer-Verlag, Berlin, Germany.

Wooldridge, M. and Jennings, N. (1999). The cooperative problem-solving process. *Journal of Logic and Computation*, **9**: 563–592.

Index